Advanced Organic Synthesis

METHODS AND TECHNIQUES

Advanced Organic Synthesis

METHODS AND TECHNIQUES

RICHARD S. MONSON

DEPARTMENT OF CHEMISTRY
CALIFORNIA STATE COLLEGE, HAYWARD
HAYWARD, CALIFORNIA

ACADEMIC PRESS
New York and London

ACADEMIC PRESS, INC.
111 Fifth Avenue, New York, New York 10003

United Kingdom Edition published by
ACADEMIC PRESS, INC. (LONDON) LTD.
Berkeley Square House, London W1X 6BA

LIBRARY OF CONGRESS CATALOG CARD NUMBER: 75-165531

PRINTED IN THE UNITED STATES OF AMERICA

Contents

4. Hydroboration

5. Catalytic Hydrogenation

6. The Introduction of Halogen

7. Miscellaneous Elimination, Substitution, and Addition Reactions

II. SKELETAL MODIFICATIONS

17.   Miscellaneous Preparations

Appendix 1.   Examples of Multistep Syntheses

Appendix 2.   Sources of Organic Reagents

Appendix 3.   Introduction to the Techniques of Synthesis

Preface

The developments in organic synthesis in recent years have been as dramatic as any that have occurred in laboratory sciences. One need only mention a few terms to understand that chemical systems that did not exist twenty years ago have become as much a part of the repertoire of the synthetic organic chemist as borosilicate glassware. The list of such terms would include the Wittig reaction, enamines, carbenes, hydride reductions, the Birch reduction, hydroboration, and so on. Surprisingly, an introduction to the manipulations of these reaction techniques for the undergraduate or graduate student has failed to materialize, and it is often necessary for students interested in organic synthesis to approach modern synthetic reactions in a haphazard manner.

The purpose of this text is to provide a survey, and systematic introduction to, the modern techniques of organic synthesis for the advanced undergraduate student or the beginning graduate student. An attempt has been made to acquaint the student with a variety of laboratory techniques as well as to introduce him to chemical reagents that require deftness and care in handling. Experiments have been drawn from the standard literature of organic synthesis including suitable modifications of several of the reliable and useful preparations that have appeared in "Organic Synthesis." Other examples have been drawn from the original literature. Where ever possible, the experiments have been adapted to the locker complement commonly found in the advanced synthesis course employing intermediate scale standard taper glassware. Special equipment for the performance of some of the syntheses would include low-pressure hydrogenation apparatus, ultraviolet lamps and reaction vessels, Dry Ice (cold finger) condensers, vacuum sublimation and distillation apparatus, and spectroscopic and chromatographic instruments. In general, an attempt has been made to employ as substrates materials that are available commercially at reasonable cost, although several of the experiments require precursor materials whose preparation is detailed in the text. Some of the reagents are hazardous to handle, but I believe that, under reasonable supervision, advanced students will be able to perform the experiments with safety.

Introductory discussion of the scope and mechanism of each reaction has been kept to a minimum. Many excellent texts and reviews exist that provide thorough and accurate discussion of the more theoretical aspects of organic synthesis, and the student is referred to these sources and to the original literature frequently. Since it is the purpose of this volume to provide technical procedures, no useful purpose would be served in merely duplicating previously explicated theoretical material.

The number of experiments that can be done satisfactorily in a one-semester course varies widely with the physical situation and the individual skills of the student. Therefore, no attempt is made to suggest a schedule. I recommend, however, that a common core of about five experiments be assigned. The remainder of the preparations can then be chosen by individual students as dictated by their interests as well as by the availability of chemicals and special equipment. The common experiments, representing frequently used and important techniques, might be chosen from Chapter 1, Sections I and IV; Chapter 2, Section I; Chapter 3, Section I; Chapter 4, Section I; Chapter 5, Section I; Chapter 6, Sections III and IV; Chapter 7, Sections II and VI; Chapter 8, Section II; Chapter 9, Sections I and II; Chapter 11, Sections I and III; or Chapter 13, Section II. Since many of the other experiments draw on the products of this suggested list, the possibility of multistep syntheses also presents itself, and several such sequences are outlined in Appendix 1. Also included, in Appendix 2, are the commercial suppliers of the chemicals required when these chemicals are not routinely available.

Finally, a brief introduction to the techniques of synthesis is given in Appendix 3. Students with no synthetic experience beyond the first-year organic chemistry course are advised to skim through this section in order to acquaint themselves with some of the apparatus and terminology used in the description of synthetic procedures.

RICHARD S. MONSON

I

FUNCTIONAL GROUP MODIFICATIONS

1

Chemical Oxidations

The controlled oxidation of selected functional groups or specific skeletal positions represents one of the most important aspects of synthetic organic chemistry. The application of a variety of inorganic oxidizing agents to organic substrates has broadened considerably the selectivity with which such oxidations may be carried out. The examples given here are typical of this rapidly expanding area of functional group modification.

I. Chromium Trioxide Oxidation

A variety of chromium (VI) oxidizing systems have been developed which allow for the oxidation of a wide range of sensitive compounds. One of the most widely used chromium (VI) reagents is the Jones reagent (1), whose use is detailed in the procedure. A related system employs acetic acid as the solvent, and an example of this reagent is also given.

A recently discovered (2) oxidizing system promises to become very important for the oxidation of acid-sensitive compounds. The reagent is chromium trioxide–pyridine complex, which may be isolated after preparation and employed in nonaqueous solvents (usually methylene chloride). A remarkable feature of the reagent is that good yields of aldehydes are obtained by direct oxidation of primary alcohols. The preparation of the reagent and its use are given.

A. Oxidation of a Commercial Mixture of *cis*- and *trans*-4-*t*-Butylcyclohexanol

$$(CH_3)_3C-\!\!\!\left\langle\;\right\rangle\!\!\!-OH \quad \xrightarrow[\text{acetone}]{H_2Cr_2O_7} \quad (CH_3)_3C-\!\!\!\left\langle\;\right\rangle\!\!\!=O$$

The chromic acid oxidizing reagent is prepared by dissolving 13.4 g of chromium trioxide in 25 ml of water. To this solution is added 12 ml of concentrated sulfuric acid. An additional minimum quantity of water is added if necessary to dissolve any precipitated salts.

A solution of 15.6 g (0.1 mole) of 4-*t*-butylcyclohexanol in 250 ml of acetone is placed in a 500-ml three-necked flask fitted with a dropping funnel, a thermometer, and a mechanical stirrer. The stirred solution is cooled in a water bath to 20°, and the chromic acid oxidizing solution is added from the dropping funnel slowly, so that the temperature of the reaction mixture does not exceed 35°. The addition is continued until the orange color of the reagent persists for about 20 minutes. The excess oxidizing agent is destroyed by the addition of a small quantity of isopropyl alcohol (1 ml is usually sufficient).

The mixture is decanted into an Erlenmeyer flask, the residual green salts are washed with two 15-ml portions of acetone, and the washings are added to the main acetone solution. Cautiously, sodium bicarbonate (approx. 13 g) is added to the solution with swirling until the pH of the reaction mixture is neutral. The suspension is filtered, and the residue is washed with 10–15 ml of acetone. The filtrate is transferred to a round-bottom flask and concentrated on a rotary evaporator under an aspirator while the flask temperature is maintained at about 50°. The flask is cooled and the residue transferred to a separatory funnel. (If solidification occurs, the residue may be dissolved in ether to effect the transfer.) To the funnel is added 100 ml of saturated sodium chloride solution, and the mixture is extracted with two 50-ml portions of ether. The ether extracts are combined, washed with several 5-ml portions of water, dried over anhydrous magnesium sulfate, and filtered into a round-bottom flask. The ether may be distilled away at atmospheric pressure (steam bath) or evaporated on a rotary evaporator. On cooling, the residue should crystallize. If it does not, it may be treated with 5 ml of 30–60° petroleum ether, and crystallization may be induced by cooling and scratching. The crystalline product is collected by filtration and recrystallized from aqueous methanol. 4-*t*-Butylcyclohexanone has mp 48–49° (yield 60–90%).

B. 4-Benzoyloxycyclohexanone from the Alcohol (*3*)

$$C_6H_5COO-\underset{}{\bigcirc}-OH \xrightarrow[\text{HOAc}]{\text{CrO}_3} C_6H_5COO-\underset{}{\bigcirc}=O$$

The oxidizing agent is prepared by dissolving 9.7 g (0.097 mole, 0.146 equivalents) of chromium trioxide in a mixture of 6 ml of water and 23 ml of acetic acid.

A 250-ml three-necked flask is equipped with a dropping funnel, a thermometer, and a mechanical stirrer, and is charged with a solution of 22 g (0.10 mole) of 4-benzoyloxycyclohexanol (Chapter 7, Section X) in 40 ml of acetic acid. The solution is cooled in a water bath, and the oxidizing solution is added at a rate so as to maintain the reaction temperature below 35°. After completion of the addition, the reaction mixture is allowed to stand at room temperature overnight. The mixture is extracted with 150 ml of ether, and the ethereal solution is washed four times with 100-ml portions of water to remove the bulk of the acetic acid. The ethereal solution is then washed with sodium bicarbonate solution followed by water and then dried over sodium sulfate. The ether is evaporated, and the residue solidifies. The product keto ester may be recrystallized from ether–petroleum ether giving plates, mp 62–63°. The yield is about 18 g (82%).

C. Chromium Trioxide–Pyridine Complex (*2, 4*)

$$CrO_3 + 2 C_5H_5N \rightarrow CrO_3(C_5H_5N)_2$$

Caution: The order of addition must be observed in this procedure or inflammation may occur.

A 1-liter flask is equipped with a magnetic stirrer, a thermometer immersed in the reaction mixture, and a drying tube. In the flask is placed 100 ml of anhydrous pyridine, and the flask is cooled in an ice-water bath to 15–20° (lower temperatures impede the complex formation). Chromium trioxide (80 g) is added in small portions to the stirred solvent at a rate so as to keep the temperature below 30°. After about one-third of the chromium trioxide has been added, the yellow complex begins to precipitate. At the end of the addition (about 1 hour), a slurry of the yellow complex in pyridine remains. (This form of the complex is apparently a microcrystalline form and is very difficult to handle.)

The temperature of the stirred solution is readjusted to 15°, and stirring at this temperature is continued until the precipitate reverts to a deep red macrocrystalline form. Petroleum ether (200 ml) is then added to the reaction mixture, the precipitate is allowed to settle, and the solvent mixture is decanted. The residue is washed three times with 200-ml portions of 30–60° petroleum ether, the solvent being removed each time by decantation. The precipitate is collected by suction filtration, dried at room temperature under a vacuum of 10 mm (higher vacuum causes some surface decomposition), and stored in a desiccator. (The complex readily forms a hydrate, which is not soluble in organic solvents. Consequently, protection from moisture is necessary.)

D. Oxidation with Chromium Trioxide–Pyridine Complex: General Procedure

$$R-CH-R(H) \xrightarrow{\text{CrO}_3(\text{C}_5\text{H}_5\text{N})_2} R-C-R(H)$$
$$\underset{OH}{|} \qquad\qquad\qquad \underset{O}{\|}$$

A 5% solution of chromium trioxide–pyridine complex in dry methylene chloride is prepared. The alcohol (0.01 mole) is dissolved in dry methylene chloride and is added in one portion to the magnetically stirred oxidizing solution (310 ml, a 6:1 mole ratio) at room temperature. The oxidation is complete in 5–15 minutes as indicated by the precipitation of the brownish black chromium reduction products. The mixture is filtered and the solvent is removed (rotary evaporator) leaving the crude product, which may be purified by distillation or recrystallization. Examples are given in Table 1.1.

II. Periodate–Permanganate Cleavage of Olefins

Oxidative cleavage of olefins is frequently a useful procedure synthetically as well as analytically. Ozonization is an effective means of carrying out such a cleavage under

TABLE 1.1

Alcohol	Product	bp (mp)/1 atm of product (°C)	(%) Yield
2-Butanol	2-Butanone	80	98
2-Octanol	2-Octanone	173	97
Cyclohexanol	Cyclohexanone	156	98
Benzhydrol	Benzophenone	(50–52)	96
1-Heptanol	Heptanal	155	93
Benzyl alcohol	Benzaldehyde	179	95
4-Nitrobenzyl alcohol	4-Nitrobenazldehyde	(105–106)	97
3-Hydroxybenzyl alcohol	3-Hydroxybenzaldehyde	(101–103)	87

mild conditions but is not always convenient. The procedure given here, discovered by Lemieux (5), employs a mixture of permanganate and periodate. The permanganate oxidizes the olefin to the 1,2-diol and the periodate cleaves the diol. The products are ketones or carboxylic acids, since any aldehydes produced are readily oxidized in the reaction medium. The troublesome accumulation of manganese dioxide is avoided in this reaction because any lower-valent derivatives of manganese that are formed are reoxidized to permanganate ion by the periodate. A good example of the use of the reaction is shown (6). The experimental procedure for the cleavage of commercial

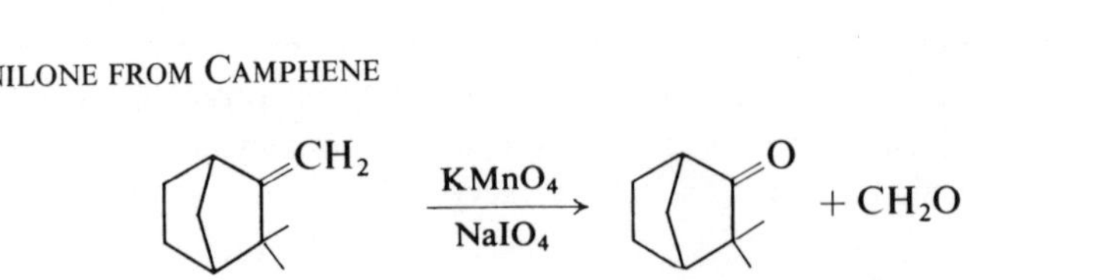

camphene is typical.

CAMPHENILONE FROM CAMPHENE

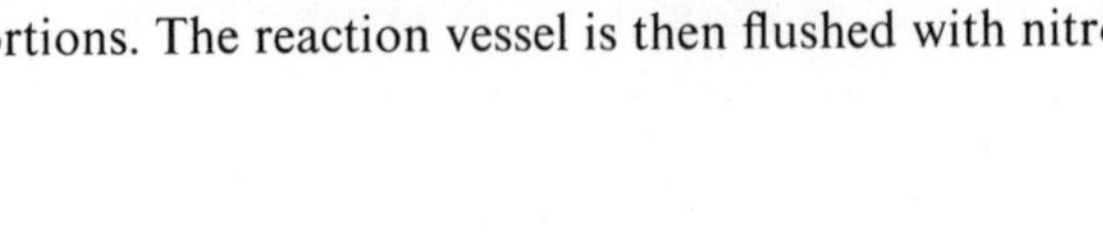

A 1-liter three-necked flask equipped with a mechanical stirrer and two dropping funnels is charged with a solution of 42 g (0.176 mole) of sodium periodate in 145 ml of acetone and 180 ml of water. To the stirred solution, camphene (6.7 g, 0.049 mole) is added in small portions. The reaction vessel is then flushed with nitrogen and is maintained under a nitrogen atmosphere thereafter. The stirred solution is cooled in an ice bath to 5°, and a solution of potassium permanganate (1.3 g, 0.008 mole) in 50 ml of water is added dropwise with the simultaneous dropwise addition of 50 ml of acetone over 1 hour, the temperature being held at 5–10°. After completion of the addition, the stirring is continued for 12 hours at 5–10°. The stirring is then discontinued, the reaction mixture is allowed to settle briefly, and the liquid reaction medium is decanted from the residue (or filtered through celite, if desired). The solution is placed on a rotary evaporator, and acetone is removed under reduced pressure. The residual aqueous phase is extracted three times with 50-ml portions of ether; the ether is washed once with saturated sodium chloride solution and dried. Removal of the ether affords the crude ketone in almost quantitative yield. It may be purified by distillation, bp 68–69°/15 mm, 193–194°/1 atm. DL-Camphenilone may be recrystallized from aqueous ethanol, mp 40°.

III. Free Radical Oxidation of an Allylic Position

The allylic position of olefins is subject to attack by free radicals with the consequent formation of stable allylic free radicals. This fact is utilized in many substitution reactions at the allylic position (cf. Chapter 6, Section III). The procedure given here employs *t*-butyl perbenzoate, which reacts with cuprous ion to liberate *t*-butoxy radical, the chain reaction initiator. The outcome of the reaction, which has general applicability, is the introduction of a benzoyloxy group in the allylic position.

3-Benzoyloxycyclohexene from Cyclohexene (7)

Caution: The reaction and the subsequent solvent removal and product distillation steps must be carried out behind a safety screen.

A 250-ml, three-necked, round-bottom flask is equipped with a mechanical stirrer, a reflux condenser, a pressure-equalizing dropping funnel, and a nitrogen inlet and outlet (mercury filled U-tube). The flask is charged with a mixture of 41 g (0.50 mole) of cyclohexene and 0.05 g of cuprous bromide, and the mixture is heated (oil bath or mantle) to 80–82°. When the temperature stabilizes, stirring is begun and 40 g (0.21 mole) of *t*-butyl perbenzoate is added dropwise over a period of about 1 hour. Stirring and heating are continued for an additional 3½ hours. The reaction mixture is cooled and washed twice with 50-ml portions of aqueous sodium carbonate to remove benzoic acid. The organic phase is washed with water until neutral and dried over anhydrous sodium sulfate. Excess cyclohexene is removed by distillation under aspirator pressure, and the residue* is distilled through a short column giving 20–33 g (71–80%) of 3-benzoyloxycyclohexene, bp 97–99°/0.15 mm, n_D^{20} 1.5376–1.5387.

IV. Epoxidation of Olefins

The reactions of olefins with peracids to form epoxides allows for the selective oxidation of carbon–carbon double bonds in the presence of other functional groups which may be subject to oxidation (for example, hydroxyl groups). The epoxides that result are easily cleaved by strong acids to diols or half-esters of diols and are therefore useful intermediates in the synthesis of polyfunctional compounds.

Caution: All reactions with organic peroxides should be conducted behind a safety shield, since peroxides occasionally explode.

A. PERBENZOIC ACID EPOXIDATION OF STYRENE (8)

$$C_6H_5CH\!=\!CH_2 + C_6H_5\underset{O}{\overset{\|}{C}}\!-\!OOH \longrightarrow C_6H_5\underset{O}{\overset{\diagdown\diagup}{CH\!-\!CH_2}} + C_6H_5COOH$$

A solution of 21 g (0.15 mole) of perbenzoic acid (Chapter 17, Section II) in 250 ml of chloroform is prepared in a 500-ml round-bottom flask. Styrene (15 g, 0.145 mole) is added, and the solution is maintained at 0° for 24 hours with frequent shaking during the first hour. At the end of the reaction period, only the slight excess of perbenzoic acid remains. The benzoic acid is extracted from the reaction mixture by washing several times with 10% sodium hydroxide solution. The solution is then washed with water and dried over anhydrous sodium sulfate. Fractional distillation gives 24–26 g (69–75%) of 1,2-epoxyethylbenzene, bp 101°/40 mm.

B. MONOPERPHTHALIC ACID EPOXIDATION OF CHOLESTERYL ACETATE (8)

* Since the perester may decompose explosively on excessive heating, an infrared spectrum of the residue should be run prior to distillation to check for complete reaction. For *t*-butyl perbenzoate, $\nu_{C=O}$ is 1775 cm^{-1} (5.63 μ).

A solution of 10 g (0.023 mole) of cholesteryl acetate (mp 112–114°) in ether (50 ml) is mixed with a solution containing 8.4 g (0.046 mole) of monoperphthalic acid (Chapter 17, Section II) in 250 ml of ether. The solution is maintained at reflux for 6 hours, following which the solvent is removed by distillation (steam bath). The residue is dried under vacuum and digested with 250 ml of dry chloroform. Filtration of the mixture gives 6.7 g of phthalic acid (87% recovery). The solvent is evaporated from the filtrate under reduced pressure and the residue is crystallized from 30 ml of methanol, giving 6.0 g (58% yield) of β-cholesteryl oxide acetate. Recrystallization affords the pure product, mp 111–112°. Concentration of the filtrate yields 1.55 g (15% yield) of α-cholesteryl oxide acetate which has a mp of 101–103° after crystallization from ethanol.

C. Hydroxylation of Cyclohexene with Hydrogen Peroxide–Formic Acid (8, 9)

$$\text{Cyclohexene} \xrightarrow[\text{HCOOH}]{\text{H}_2\text{O}_2} \text{(epoxide)} \xrightarrow[\text{H}_2\text{O}]{\text{H}^+} \text{1,2-cyclohexanediol}$$

Cyclohexene (8 g, 0.097 mole) is added to a mixture of 105 g of 98–100% formic acid and 13 g (0.115 mole) of 30% hydrogen peroxide contained in a 250-ml flask fitted with a reflux condenser. The two layers are shaken together briefly, whereupon spontaneous heating occurs. The mixture becomes homogeneous at 65–70°, this temperature being maintained for 2 hours on a steam bath. The formic acid is removed by distillation under reduced pressure. The residue is mixed with 50 ml of 6 N sodium hydroxide and heated on a steam bath for 45 minutes. The solution is cooled, neutralized with hydrochloric acid and evaporated to dryness under vacuum. The solid residue is distilled, affording about 10 g of the product, bp 128–132°/15 mm. The distillate solidifies and may be recrystallized from acetone, giving about 70% of trans-1,2-cyclohexanediol, mp 102–103°.

V. Baeyer-Villiger Oxidation of Ketones

The reaction of peracids with ketones proceeds relatively slowly but allows for the conversion of ketones to esters in good yield. In particular, the conversion of cyclic ketones to lactones is synthetically useful because only a single product is to be expected. The reaction has been carried out with both percarboxylic acids and Caro's acid (formed by the combination of potassium persulfate with sulfuric acid). Examples of both procedures are given.

A. Perbenzoic Acid Oxidation of Ketones: General Procedure (10)

$$\underset{\underset{\text{O}}{\|}}{\text{RCCH}_3} + \underset{\underset{\text{O}}{\|}}{\text{C}_6\text{H}_5\text{COOH}} \xrightarrow{\text{benzene}} \underset{\underset{\text{O}}{\|}}{\text{R—OCCH}_3} + \text{C}_6\text{H}_5\text{COOH}$$

A perbenzoic acid solution in benzene is prepared as in Chapter 17, Section II. (This solution is approximately 1.8 M in perbenzoic acid.) To 67 ml (approx. 0.12 mole of perbenzoic acid) of this solution contained in an Erlenmeyer flask is added 0.10 mole of the ketone in one batch. The resulting solution is swirled at intervals and allowed to stand at room temperature for 10 days. The solution is then washed three times with 50-ml portions of saturated sodium bicarbonate solution to remove benzoic acid and unreacted peracid, and is then washed with water. The solution is dried (anhydrous sodium sulfate), the benzene is evaporated, and the residue is fractionally distilled at reduced pressure to give the ester.

Examples

1. Cyclohexyl methyl ketone gives cyclohexyl acetate, bp 74–77°/23 mm.
2. Cyclohexanone gives ε-caprolactone, bp 102–104°/7 mm, which may polymerize on standing. The lactone may be converted easily to the corresponding ε-hydroxyhydrazide by heating on a steam bath with a slight excess of 100% hydrazine hydrate. The crude hydrazide may be recrystallized from ethyl acetate, mp 114–115°.

PERSULFATE OXIDATION OF KETONES: GENERAL PROCEDURE (*11*)

$$\underset{\underset{O}{\|}}{R-C-CH_3} \quad \xrightarrow[\text{H}_2\text{SO}_4]{\text{K}_2\text{S}_2\text{O}_8} \quad \underset{\underset{O}{\|}}{R-O-C-CH_3}$$

The oxidizing agent is prepared in a 500-ml flask equipped with a magnetic stirrer and cooled in an ice bath as follows: In the flask are placed 60 ml of concentrated sulfuric acid and 20 ml of water, and the solution is cooled to 10°. Potassium persulfate (42 g, 0.15 mole) is added slowly to the stirred solution while maintaining the temperature below 10°. The solution is diluted with an additional 65 ml of water maintaining the temperature below 15°. The solution is now cooled to about 7° and 0.08 mole of the ketone is added over 40 minutes. After the addition has been completed, the solution is allowed to come to room temperature and stirring is continued for 20 hours. The solution is diluted carefully with 150 ml of water and extracted twice with 75-ml portions of ether. The ether is washed with sodium bicarbonate solution, followed by water, and the ethereal solution is dried. Removal of the solvent, followed by fractional distillation, affords the product ester.

Examples

1. Cyclopentanone gives δ-valerolactone, bp 98–100°/5 mm. The lactone may be converted as above to its δ-hydroxyhydrazide and recrystallized from ethyl acetate, mp 105–106°.
2. 2-Methylcyclopentanone gives δ-methyl-δ-valerolactone, bp 100–101°/5 mm.

VI. Lead Tetraacetate Oxidation of Cycloalkanols

The reaction of lead tetraacetate (LTA) with monohydric alcohols produces functionalization at a remote site yielding derivatives of tetrahydrofuran (THF) (*12*). An example is the reaction of 1-pentanol with LTA in nonpolar solvents which produces 30% THF. The reaction, which is believed to proceed through free-radical intermediates, gives a variable distribution of oxidation products depending on solvent polarity, temperature, reaction time, reagent ratios, and potential angle strain in the product.

In the procedure given here, the reaction is applied to a cyclic alcohol to produce a bridged ether. The product is of interest in that it can be cleaved to produce disubstituted cyclooctanes of known geometry (cf. Chapter 6, Section V).

LEAD TETRAACETATE OXIDATION OF CYCLOOCTANOL (*13*)

A mixture of 200 ml dry benzene, 35.5 g (0.08 mole) of lead tetraacetate, and 15 g of anhydrous calcium carbonate is placed in a 500-ml round-bottom flask equipped with a magnetic stirrer, a heating mantle, and a condenser (drying tube). The mixture is heated with stirring to reflux. Ten grams (0.078 moles) of cyclooctanol dissolved in 50 ml of dry benzene are added in one batch, and the mixture is refluxed for 48 hours. The mixture is then cooled and filtered by suction through celite. Water (25 ml) is added to the filtrate, and the solution is stirred for $\frac{1}{2}$ hour to hydrolize any unreacted lead tetraacetate. The mixture is then filtered again by suction through celite giving a clear two-phase filtrate. The water layer is separated, and the benzene layer is dried over anhydrous sodium sulfate, filtered, and fractionally distilled. The major product is the cyclic oxide which has bp 50–52°/6.5 mm and mp 30–32°.

VII. Photolytic Conversion of Cyclohexane to Cyclohexanone Oxime

Nitrosyl chloride reacts with aliphatic hydrocarbons at room temperature under the influence of light to give a complex mixture of substitution products. When the reaction is run on cyclohexane at −25°, however, the pure oxime hydrochloride crystallizes from the reaction mixture with virtually no side products.

A. Cyclohexanone Oxime from Cyclohexane (14)

$$\text{(cyclohexane)} + NOCl \xrightarrow{h\nu} \text{(cyclohexanone oxime)} {=}NOH \cdot HCl$$

Caution: Nitrosyl chloride is an extremely corrosive gas, and all operations with it should be carried out in a hood.

A 500-ml three-necked flask is fitted with a gas inlet tube, a magnetic stirrer, and an alcohol thermometer, and is immersed in a Dry Ice–acetone bath to a depth sufficient to maintain an internal temperature of -20 to $-25°$. Four 150-watt spotlights are mounted around the flask. The cooling bath and the gas inlet tube should be protected from the light by covering with aluminum foil. A cyclohexane–benzene mixture (70:30 by volume) is placed in the flask to a convenient volume (100–200 ml) and is allowed to cool to $-20°$. The lamps are turned on, and nitrosyl chloride is added slowly to the flask through the immersed gas inlet tube. The solution gets cloudy in about 10 minutes and cyclohexanone oxime hydrochloride precipitates in 15–20 minutes. The addition of a total of 7 g of nitrosyl chloride should take about 5 hours. Under these circumstances about 9 g of the oxime hydrochloride can be isolated by filtration followed by washing with ether and drying in a vacuum oven or desiccator. The oxime hydrochloride has mp 70–88° and is quite hygroscopic. It may be converted to the free oxime by the following procedure: The oxime hydrochloride (8 g) is suspended in 180 ml of boiling dry ether. Dry ammonia is bubbled slowly through the mixture for several hours. The ammonium chloride formed is filtered off, and the ether is removed from the filtrate on a rotary evaporator. This residue is recrystallized from petroleum ether. The free oxime has mp 88–90°.

VIII. Oxidation of Ethers to Esters

The oxidation of ethers to esters according to the reaction offers many possibilities for the modification of functionality in open chain or cyclic systems. An example is the

$$R-CH_2-O-R' \longrightarrow R-\underset{\underset{O}{\|}}{C}-O-R'$$

conversion of tetrahydrofurans to γ-butyrolactones. Two reagents have been discovered that allow for this conversion in satisfactory yield: ruthenium tetroxide and trichloroisocyanuric acid (Chapter 17, Section IV). The use of these reagents is given below for the conversion of di-*n*-butyl ether to *n*-butyl *n*-butyrate.

A. *n*-Butyl Butyrate from Di-*n*-butyl Ether by Ruthenium Tetroxide

$$(CH_3CH_2CH_2CH_2)_2O \xrightarrow{RuO_4} CH_3CH_2CH_2\underset{\underset{O}{\|}}{C}OCH_2CH_2CH_2CH_3$$

1. *Preparation of Ruthenium Tetroxide* (*15*): In a 250-ml flask equipped with a magnetic stirrer and cooled in an ice–salt bath is placed a mixture of 0.4 g of ruthenium dioxide and 50 ml of carbon tetrachloride. A solution of 3.2 g of sodium metaperiodate in 50 ml of water is added and the mixture is stirred 1 hour at 0°. The black ruthenium dioxide gradually dissolves. The clear yellow carbon tetrachloride layer is separated and filtered through glass wool to remove insoluble materials. The solution may be used immediately or stored in the cold in the presence of 50 ml of sodium metaperiodate solution (1 g/50 ml). As prepared above, the solution is about 0.037 M in ruthenium tetroxide and contains 0.3 g/50 ml.

2. *Oxidation of Di-n-butyl Ether* (*16*): The ruthenium tetroxide solution (containing about 0.3 g of the oxidizing agent) is added dropwise to a magnetically stirred solution of 0.40 g of di-*n*-butyl ether in 10 ml of carbon tetrachloride cooled in an ice bath. A thermometer is inserted in the reaction mixture. After a few minutes, black ruthenium dioxide begins to form, and the temperature rises. The rate of addition is controlled to maintain the temperature at 10–15°. After completion of the addition, the reaction mixture is allowed to stand at room temperature overnight. The precipitated ruthenium dioxide is filtered off, and the residue is washed thoroughly with carbon tetrachloride. The combined filtrate and washings are washed once with sodium bicarbonate solution to remove a trace of butyric acid. The carbon tetrachloride solution is then dried (anhydrous sodium sulfate), filtered, and distilled in a micro-apparatus. *n*-Butyl *n*-butyrate has a normal boiling point of 165–166°.

B. n-BUTYL BUTYRATE FROM DI-n-BUTYL ETHER BY TRICHLOROISOCYANURIC ACID (*17*)

In a 200-ml round-bottom flask equipped with a magnetic stirrer and a thermometer is placed a mixture of 50 ml of di-*n*-butyl ether and 25 ml of water. The flask is immersed in an ice bath and the mixture is cooled to 5°. In one portion is added 23.2 g (0.1 moles) of trichloroisocyanuric acid (Chapter 17, Section IV), and stirring in the ice bath is continued for 12 hours. The ice bath is removed and the mixture is stirred at room temperature for an additional 8 hours. The reaction mixture is then filtered to remove solids. The water is separated from the organic layer, which is then washed with two additional portions of water, dried with anhydrous sodium sulfate, filtered, and fractionated as above.

IX. Partial Oxidation of an Aliphatic Side Chain

As mentioned earlier, allylic positions are highly subject to attack by free radicals. Likewise, benzylic positions may be attacked by free-radical initiating reagents to give benzylic radicals of high stability. The procedure given below employs cerium (IV) in conjunction with nitric acid to carry out the oxidation of the benzylic position of Tetralin. Although cerium (IV) reagents have been widely used in inorganic analytic procedures, their use in organic oxidations is relatively recent.

α-Tetralone from Tetralin (*18*)

A three-necked round-bottom flask is fitted with a dropping funnel, a thermometer, and a magnetic stirrer and is heated in a water bath to 30°. Tetralin (1.32 g, 0.01 mole) and 50 ml of 3.5 *N* nitric acid solution are placed in the flask and brought to temperature. Ceric ammonium nitrate (21.9 g, 0.04 mole) is dissolved in 100 ml of 3.5 *N* nitric acid, and the solution is added dropwise to the reaction mixture at a rate such that the temperature does not rise and only a pale yellow color is evident in the reaction mixture. At the completion of the reaction (1½ to 2 hours), the mixture should be colorless. The solution is cooled to room temperature, diluted with an equal volume of water, and extracted twice with ether. The ether solution is dried with anhydrous sodium sulfate, filtered, and the ether is evaporated. The residue may be distilled to yield α-tetralone (bp 113–116°/6 mm or 170°/49 mm) or may be converted directly to the oxime, which is recrystallized from methanol, mp 88–89°.

X. Bisdecarboxylation with Lead Tetraacetate

The use of lead tetraacetate to carry out oxidative bisdecarboxylation of diacids has been found to be a highly useful procedure when used in conjunction with the Diels-Alder addition of maleic anhydride to dienes, the latter process providing a ready source of 1,2-dicarboxylic acids. The general pattern is illustrated in the reaction

sequence. A striking use of the reaction was made by van Tamelen and Pappas (*19*) in their synthesis of Dewar benzene.

General Procedure (20)

Pyridine (purified by distillation from barium oxide, 10 ml/g of diacid) is placed in a round-bottom flask fitted with a magnetic stirrer, condenser, and drying tube. Oxygen is bubbled through the stirred solution at room temperature for 15 minutes. The diacid (0.02 mole) and lead tetraacetate (0.03 mole) are added, and the flask is heated with a mantle or oil bath to 65° while the stirring is continued. The evolution of carbon dioxide begins after several minutes and is usually complete after an additional 10 minutes. The reaction mixture is cooled, poured into excess dilute nitric acid, and extracted with ether. The ether solution is washed with aqueous bicarbonate then saturated sodium chloride solution and finally dried over anhydrous magnesium sulfate. The solution is filtered, and the solvent is removed by a rotary evaporator or by fractionation to give the olefin.

Examples (20)

1. *cis*-4,5-Cyclohexenedicarboxylic acid (Chapter 8, Section II) is converted to 1,4-cyclohexadiene, bp 86–87°, n_D^{20} 1.4729.

2. *endo*-1-Acetoxy-8,8-dimethylbicyclo[2.2.2]oct-3-one-5,6-dicarboxylic acid gives the corresponding olefin, 1-acetoxy-8,8-dimethylbicyclo[2.2.2]oct-2-ene-5-one, mp 59–60° after recrystallization from pentane (Chapter 8, Section IV).

3. *endo*-1-Acetoxybicyclo[2.2.2]oct-3-one-5,6-dicarboxylic acid gives the corresponding olefin, 1-acetoxybicyclo[2.2.2]oct-2-en-5-one, mp 49–50° after recrystallization from pentane (Chapter 8, Section IV).

XI. Oxidation with Selenium Dioxide

Selenium dioxide may be used for the oxidation of reactive methylene groups to carbonyl groups. An example is the oxidation of cyclohexanone to cyclohexane-1,2-dione (21). In the procedure, the reaction is carried out on camphor to give camphor

quinone, an intermediate in the preparation of Horner's Acid (see Chapter 15, Section II).

CAMPHOR QUINONE (*22*)

Caution: Selenium dioxide is extremely poisonous and proper care should be taken when working with it.

Commercial selenium dioxide gives more consistent results when freshly sublimed material is used. Place the oxide (50 g) in a 7-cm porcelain crucible upon which is set a 250-ml filter flask cooled by running through it a stream of water. The crucible is heated with a low flame until sublimation is complete (20–30 minutes). After cooling, the sublimed selenium dioxide is scraped from the flask and is stored in a stoppered bottle.

In a 100-ml flask is placed a mixture of 19.5 g (0.18 mole) of freshly sublimed, pulverized selenium dioxide, 15 g (0.10 mole) of *dl*-camphor and 15 ml of acetic anhydride. The flask is fitted with a magnetic stirrer and a condenser, and the mixture is heated to 135° on an oil bath with stirring for 16 hours. After cooling, the mixture is diluted with ether to precipitate selenium, which is then filtered off, and the volatile materials are removed under reduced pressure. The residue is dissolved in ether (200 ml), washed four times with 50-ml portions of water and then washed several times with saturated sodium bicarbonate solution (until the washes are basic). The ether solution is finally washed several times with water, then dried, and the ether is evaporated. The residue may be purified by sublimation at reduced pressure or recrystallized from aqueous ethanol (with clarification by Norit, if necessary). The product is yellow, mp 197–199°.

REFERENCES

1. A. Bowers, T. G. Halsall, E. R. H. Jones, and A. J. Lemin, *J. Chem. Soc.*, p. 2548 (1953); E. J. Eisenbraun, *Org. Syn.* **45**, 28 (1965); K. B. Wiberg, ed., "Oxidation in Organic Chemistry." Academic Press, New York, 1965.
2. J. C. Collins, W. W. Hess, and F. J. Frank, *Tetrahedron Lett.*, p. 3363 (1968).
3. E. R. H. Jones and F. Sondheimer, *J. Chem. Soc.*, p. 615 (1949).
4. G. I. Poos, G. E. Arth, R. E. Beyler, and L. H. Sarett, *J. Amer. Chem. Soc.* **75**, 422 (1953); J. R. Holum, *J. Org. Chem.* **26**, 4814 (1961).
5. R. U. Lemieux and E. von Rudloff, *Can. J. Chem.* **33**, 1701, 1710, 1714 (1955); *Can. J. Chem.* **34**, 1413 (1956).
6. C. G. Overberger and H. Kay, *J. Amer. Chem. Soc.* **89**, 5640 (1967).
7. K. Pedersen, P. Jakobsen, and S. Lawesson, *Org. Syn.* **48**, 18 (1968) and references cited therein.
8. D. Swern, *Org. React.* **7**, 378 (1953).
9. A. Roebuck and H. Adkins, *Org. Syn. Collective Vol.* **4**, 217 (1955).
10. S. L. Friess, *J. Amer. Chem. Soc.* **71**, 14, 2571 (1949).
11. T. H. Parliment, M. W. Parliment, and I. S. Fagerson, *Chem. Ind.* (*London*), p. 1845 (1966).

12. R. E. Partch, *J. Org. Chem.* **30**, 2498 (1965); K. Heusler and K. Kalvoda, *Angew. Chem. Int. Ed. Engl.* **3**, 525 (1964).
13. R. M. Moriarty and H. G. Walsh, *Tetrahedron Lett.*, p. 465 (1965).
14. M. A. Naylor and A. W. Anderson, *J. Org. Chem.* **18**, 115 (1953).
15. H. Nakata, *Tetrahedron* **19**, 1959 (1963).
16. L. M. Berkowitz and P. N. Rylander, *J. Amer. Chem. Soc.* **80**, 6682 (1958).
17. E. C. Juenge and D. A. Beal, *Tetrahedron Lett.*, p. 5819 (1968).
18. L. Syper, *Tetrahedron Lett.*, p. 4493 (1966).
19. E. E. van Tamelen and S. P. Pappas, *J. Amer. Chem. Soc.* **85**, 3297 (1963).
20. C. M. Cimarusti and J. Wolinsky, *J. Amer. Chem. Soc.* **90**, 113 (1968).
21. C. C. Hach, C. V. Banks, and H. Diehl, *Org. Syn. Collective Vol.* **4**, 229 (1963).
22. R. R. Pennelly and J. C. Shelton, private communication.

2

Hydride and Related Reductions

The availability of a wide range of complex hydride reducing agents has greatly simplified the problem of selective reduction of functional groups. Virtually any polar functional group can be reduced by appropriate selection of a hydride source. Another important feature of hydride reductions is the extent to which the geometry of the product can be controlled. By appropriate choice of conditions, pure axial or equatorial isomers of cyclohexyl derivatives can be prepared. The procedures given here entail some of the general features of hydride reduction as well as some of the stereospecific modifications (*1*).

I. Reduction by Lithium Aluminum Hydride

The most versatile of the complex hydrides is lithium aluminum hydride. It is sufficiently reactive to reduce carboxylate ions directly to primary alcohols and is of course useful for the reduction of less sluggish substrates. This high reactivity, however, means that care must be exercised in the handling of the reagent. It reacts rapidly with moisture or protic solvents, resulting in the liberation of hydrogen and the attendant danger of explosion, and direct contact with liquid water may cause ignition. Vigorous grinding in the presence of air may also cause fire, and sharp mechanical shock has been known to cause explosion. But carefully dried solvents are not required, since the use of a slight excess of the reagent serves as a convenient and effective drying agent. The procedures below describe the use of the reagent for the reduction of a diester, a hindered carboxylic acid, and a substituted amide.

A. 1,6-Hexanediol from Diethyl Adipate (*2*)

$$C_2H_5OOC-(CH_2)_4-COOC_2H_5 \xrightarrow[\text{ether}]{\text{LiAlH}_4} HO-(CH_2)_6-OH$$

A three-necked, round-bottom, 500-ml flask is fitted with a mechanical stirrer, a dropping funnel, and a condenser with openings protected by drying tubes. Lithium aluminum hydride (3.5 g) is placed in the flask with 100 ml of anhydrous ether, and stirring is begun. After 10 minutes, a solution of 16.5 g of diethyl adipate in 50 ml of anhydrous ether is added dropwise at such a rate that a gentle ether reflux is maintained. If the reaction mixture becomes viscous, 15-ml portions of anhydrous ether may be added as necessary to facilitate stirring. After the completion of the addition, stirring is continued for 10 minutes. Excess hydride should then be decomposed by addition of 7–8 g of ethyl acetate with stirring. The solution is decanted from the sludge, and the sludge is dissolved in 3 M sulfuric acid solution. The solution is extracted three times with 50-ml portions of ether and the extracts are combined with the original ether solution. The solution is dried over anhydrous sodium sulfate and filtered, and the ether is evaporated. The residue solidifies and may be recrystallized from water. 1,6-Hexanediol has mp 41–42°.

B. *trans*-9-DECALYLCARBINOL FROM THE ACID (3)

A setup similar to the preceding one is used in this experiment except that provision should be made for heating the reaction vessel (steam bath, oil bath, or mantle). Lithium aluminum hydride (10 g, 0.26 mole) is dissolved in 200 ml of dry *n*-butyl ether and heated with stirring to 100°. A solution of 9.1 g (0.05 mole) of *trans*-9-decalin-carboxylic acid (Chapter 16, Section I) in 100 ml of dry *n*-butyl ether is added dropwise over about 30 minutes. The stirring and heating are continued for 4 days, after which the mixture is cooled and water is slowly added to decompose excess hydride. Dilute hydrochloric acid is added to dissolve the salts, and the ether layer is separated, washed with bicarbonate solution then water, and dried. The solvent is removed by distillation, and the residue is recrystallized from aqueous ethanol, mp 77–78°, yield 80–95%.

C. N,N-DIMETHYLCYCLOHEXYLMETHYLAMINE FROM CYCLOHEXANECARBOXYLIC ACID (4)

1. *N,N-Dimethylcyclohexanecarboxamide:* A 500-ml three-necked flask is equipped with a reflux condenser (drying tube), a pressure-equalizing dropping funnel, and a magnetic stirrer. The flask is charged with 32 g (0.25 mole) of cyclohexanecarboxylic acid and thionyl chloride (45 g, 0.375 mole) is added over 5 minutes to the stirred acid. The flask is heated (oil bath) at a temperature of 150° for 1 hour. The reflux condenser is

then replaced by a distillation head, 50 ml of anhydrous benzene is added, and the mixture is distilled until the temperature of the vapors reaches 95°. The mixture is cooled, another 50 ml of anhydrous benzene is added, and the distillation process is repeated to a head temperature of 95°. The cooled acid chloride is now transferred with a little benzene to a dropping funnel attached to a 500-ml three-necked flask. The flask is fitted with a mechanical stirrer and a drying tube and is immersed in an ice bath. A solution of 34 g (0.75 mole) of anhydrous dimethylamine in 40 ml of anhydrous benzene is placed in the flask. The acid chloride is added slowly to the stirred solution, the addition taking about 1 hour. The mixture is then stirred at room temperature overnight. Water (50 ml) is added, the layers are separated, and the aqueous phase is extracted with two 25-ml portions of ether. The combined extracts and benzene layer are washed with saturated sodium chloride solution and dried over anhydrous magnesium sulfate. The solvent is removed (rotary evaporator) and the residue is distilled under reduced pressure. The yield of N,N-dimethylcyclohexanecarboxamide is 33–35 g (86–89%), n_D^{25} 1.4800–1.4807, bp 85–86°/1.5 mm.

2. *N,N-Dimethylcyclohexylmethylamine:* A 500-ml three-necked flask fitted with a reflux condenser (drying tube), a pressure-equalizing dropping funnel, and a magnetic stirrer is charged with a mixture of 5 g of lithium aluminum hydride in 60 ml of anhydrous ether. A solution of 20 g of N,N-dimethylcyclohexanecarboxamide in 50 ml of anhydrous ether is added to the stirred mixture at a rate so as to maintain a gentle reflux (about 30 minutes). The mixture is then stirred and heated (mantle) at reflux for 15 hours. The flask is fitted with a mechanical stirrer and cooled in an ice bath. Water (12 ml) is added slowly with vigorous stirring, and the stirring is continued for an additional 30 minutes. A cold solution of 30 g of sodium hydroxide in 75 ml of water is added and the mixture is steam-distilled until the distillate is neutral (about 225 ml of distillate). The distillate is acidified by addition of concentrated hydrochloric acid (approx. 15 ml) with cooling. The layers are separated, and the ether layer is washed with 10 ml of 3 N hydrochloric acid. The combined acidic solutions are concentrated at 20 mm pressure until no more distillate comes over at steam-bath temperature. The residue is dissolved in water (approx. 30 ml), the solution is cooled, and sodium hydroxide pellets (17 g) are added slowly, with stirring and cooling in an ice-water bath. The layers are separated, and the aqueous phase is extracted with three 15-ml portions of ether. The combined organic layer and ether extracts are dried over potassium hydroxide pellets for 3 hours. The solvent is removed by fractional distillation at atmospheric pressure, and the product is collected by distillation under reduced pressure. The yield is about 16 g (88%) of N,N-dimethylcyclohexylmethylamine, bp 76°/29 mm, n_D^{25} 1.4462–1.4463.

II. Mixed Hydride Reduction (5)

The lithium aluminum hydride–aluminum chloride reduction of ketones is closely related mechanistically to the Meerwein-Ponndorf-Verley reduction in that the initially formed alkoxide complex is allowed to equilibrate between isomers in the presence of a catalytic amount of ketone via an intermolecular hydride ion transfer. If the energy difference between the isomers is sufficiently great, virtually all the complex is present as the stable isomer at equilibrium. This fact is applied in the procedure to the reduction products of 4-*t*-butylcyclohexanone. Thus, when the complex is finally decomposed, the pure *trans*-4-*t*-butylcyclohexanol is easily isolated.

A second example exploits the fact that the mixed hydride reagent is capable of hydrogenolysis of certain carbon-oxygen bonds. Thus, treatment of cyclohexanone ketal (Chapter 7, Section IX) with lithium aluminum hydride–aluminum chloride results in the rupture of a C–O bond to give the oxyethanol derivative.

A. *trans*-4-*t*-BUTYLCYCLOHEXANOL FROM THE KETONE (6)

In a 500-ml three-necked flask, equipped with a mechanical stirrer, a dropping funnel, and a reflux condenser (drying tube), is placed 6.7 g (0.05 mole) of anhydrous aluminum chloride. The flask is cooled in an ice bath, 50 ml of dry ether is slowly added from the dropping funnel, and the mixture is stirred briefly. Powdered lithium aluminum hydride (0.6 g) is placed in a 100-ml flask fitted with a condenser, and 20 ml of dry ether is added slowly from the top of the condenser while the flask is cooled in an ice bath. The mixture is refluxed for 30 minutes then cooled, and the resulting slurry is transferred to the dropping funnel on the 500-ml flask. The slurry is added to the stirred ethereal solution of aluminum chloride over 10 minutes, and the reaction mixture is stirred for an additional 30 minutes without cooling to complete the formation of the "mixed hydride".

The dropping funnel is charged with a solution of 7.7 g (0.05 mole) of 4-*t*-butylcyclo-hexanone (Chapter 1, Section I) in 50 ml of dry ether. The solution is slowly added to the "mixed hydride" solution at a rate so as to maintain a gentle reflux. The reaction mixture is then refluxed for an additional 2 hours. Excess hydride is consumed by the addition of 1 ml of dry *t*-butyl alcohol, and the mixture is refluxed for 30 minutes more. 4-*t*-Butylcyclohexanone (0.3 g) in 5 ml of dry ether is added to the reaction mixture, and refluxing is continued for 4 hours. The cooled (ice bath) reaction mixture is decomposed by the addition of 10 ml of water followed by 25 ml of 10% aqueous sulfuric acid. The ether layer is separated, and the aqueous layer is extracted with 20 ml of ether. The combined ether extracts are washed with water and dried over anhydrous magnesium sulfate. After filtration, the ether is removed (rotary evaporator), and the residue

(7–8 g) solidifies. Analysis by gas-liquid phase chromatography (glpc) shows it to contain 96% trans alcohol, 0.8% cis alcohol and 3.2% ketone. Recrystallization of the crude product from 60–90° petroleum ether gives 5–6 g of the product, mp 75–78°, whose approximate composition is 99.3% trans alcohol, 0.3% cis alcohol, and 0.4% ketone. Concentration of the mother liquor affords a second crop, which is sufficiently pure for most preparative purposes.

B. 2-CYCLOHEXYLOXYETHANOL (7)

As in the preceding experiment, a lithium aluminum hydride–aluminum chloride reducing agent is prepared by the addition of 1.67 g (0.044 mole) of lithium aluminum hydride in 50 ml of anhydrous ether to 24.2 g (0.18 mole) of anhydrous aluminum chloride in 50–55 ml of anhydrous ether.

A solution of 12.5 g (0.088 mole) of 1,4-dioxaspiro[4.5]decane (Chapter 7, Section IX) in 200 ml of anhydrous ether is added to the stirred mixture at a rate so as to maintain a gentle reflux. (Cooling in an ice bath is advisable.) The reaction mixture is then refluxed for 3 hours on a steam bath. Excess hydride is carefully destroyed by the dropwise addition of water (1–2 ml) to the ice-cooled vessel until hydrogen is no longer evolved. Sulfuric acid (100 ml of 10% solution) is now added followed by 40 ml of water, resulting in the formation of two clear layers. The ether layer is separated and the aqueous layer extracted with three 20-ml portions of ether. The combined ethereal extracts are washed with saturated sodium bicarbonate solution followed by saturated sodium chloride solution. The ethereal solution is dried over anhydrous potassium carbonate (20–24 hours), filtered, and concentrated by distillation at atmospheric pressure. The residue is distilled under reduced pressure affording 2-cyclohexyloxy-ethanol as a colorless liquid, bp 96–98°/13 mm, n_D^{25} 1.4600–1.4610, in about 85% yield.

III. Reduction with Iridium-Containing Catalysts

A recently discovered reduction procedure provides a convenient route to axial alcohols in cyclohexyl derivatives (8). The detailed mechanism of the reaction remains to be elucidated, but undoubtedly the reducing agent is an iridium species containing one or more phosphate groups as ligands. In any case, it is clear that the steric demands of the reducing agent must be extraordinary since the stereochemical outcome of the reaction is so specific. The procedure below is for the preparation of a pure axial alcohol from the ketone.

cis-4-*t*-Butylcyclohexanol from the Ketone (*9*)

To a solution of 1.0 g (0.003 mole) of iridium tetrachloride in 0.5 ml of concentrated hydrochloric acid is added 15 ml of trimethylphosphite. This solution is added to a solution of 7.7 g (0.05 mole) of 4-*t*-butylcyclohexanone in 160 ml of isopropanol in a 500-ml flask equipped with a reflux condenser. The solution is refluxed for 48 hours, then cooled, and the isopropanol is removed on a rotary evaporator. The residue is diluted with 65 ml of water and extracted four times with 40-ml portions of ether. The extracts are dried with anhydrous magnesium sulfate, filtered, and the ether is removed on the rotary evaporator. The white solid residue is recrystallized from 60% aqueous ethanol affording cis alcohol of greater than 99% purity, mp 82–83.5°.

IV. Reduction of Conjugated Alkenes with Chromium (II) Sulfate

Chromium (II) sulfate is capable of reducing a variety of functional groups under mild conditions (*10*). Of particular interest is its ability to reduce α,β-unsaturated esters, acids, and nitriles to the corresponding saturated compounds. This capability is illustrated in the procedure by the reduction of diethyl fumarate.

A. Chromium (II) Sulfate Solution (*10*)

$$Cr_2(SO_4)_3 + Zn(Hg) \rightarrow 2\,CrSO_4 + ZnSO_4$$

A 500-ml three-necked flask fitted with a mechanical stirrer and a nitrogen inlet and outlet is charged with 30 g (approx. 0.055 mole) of hydrated chromium (III) sulfate, 200 ml of distilled water, 7.5 g (0.12 g-atom) of mossy zinc, and 0.4 ml (5.4 g, 0.03 g-atom) of mercury. The flask is flushed with nitrogen for 30 minutes and a nitrogen atmosphere is maintained. The mixture is then heated to about 80° with stirring for 30 minutes to initiate reaction. Then the mixture is stirred at room temperature for an additional 30 hours, by which time the green mixture has been converted to a clear blue solution. Solutions prepared as above are about 0.55 M in chromium (II) and are indefinitely stable if protected from oxygen.

B. Reduction of Diethyl Fumarate (*10*)

A nitrogen atmosphere is maintained over the reaction mixture prepared above. The flask is fitted with a pressure-equalizing addition funnel containing a solution of 8.7 g (0.05 mole) of diethyl fumarate in 85 ml of dimethylformamide (DMF). With stirring, the diethyl fumarate is added rapidly. The solution immediately turns green, and the reduction is complete in 10 minutes. The resulting solution is diluted with 65 ml of water, ammonium sulfate (20 g) is added, and the mixture is extracted with four 100-ml portions of ether. The combined extracts are washed three times with 30-ml portions of water and dried over anhydrous magnesium sulfate. The ether is removed (rotary evaporator) and the residual liquid is distilled affording diethyl succinate, bp 129°/44 mm, n_D^{25} 1.4194, in about 90% yield.

REFERENCES

1. N. G. Gaylord, "Reduction with Complex Metal Hydrides." Interscience, New York, 1956; W. G. Brown, *Org. React.* **6**, 469 (1951).
2. A. I. Vogel, "Practical Organic Chemistry," 3rd ed. Longmans, London, 1956.
3. W. G. Dauben, R. C. Tweit, and R. L. MacLean, *J. Amer. Chem. Soc.* **77**, 48 (1955).
4. A. C. Cope and E. Ciganek, *Org. Syn. Collective Vol.* **4**, 339 (1963).
5. E. L. Eliel, *Rec. Chem. Progr.* **22**, 129 (1961).
6. E. L. Eliel, R. J. L. Martin, and D. Nasipuri, *Org. Syn.* **47**, 16 (1967).
7. R. A. Daignault and E. L. Eliel, *Org. Syn.* **47**, 37 (1967).
8. Y. M. Y. Haddad, H. B. Henbest, J. Husbands, and T. R. B. Mitchell, *Proc. Chem. Soc. London*, p. 361 (1964).
9. E. L. Eliel, T. W. Doyle, R. O. Hutchins, and E. C. Gilbert, cited in M. Fieser and L. Fieser, "Reagents for Organic Synthesis," Vol. 2, p. 228. Wiley/Interscience, New York, 1969.
10. A. Zurqiyah and C. E. Castro, *Org. Syn.* **49**, 98 (1969) and references cited therein.

3

Dissolving Metal Reductions

Although once used extensively for the reduction of functional groups, reactions employing dissolving metals have largely been replaced by other more convenient methods. Nevertheless, certain synthetic sequences that may require stereospecific or functionally selective reductions may best be executed by means of metals in solution. The Birch reduction, or its modifications (*1*), employs a solution of an alkali metal in liquid ammonia or an aliphatic amine and is still widely used in connection with the reduction of aromatic or conjugated systems. The sequence showing the reduction of benzene to 1,4-cyclohexadiene is typical of the one-electron transfer mechanism commonly understood to pertain in such reductions. The procedures given below exemplify

the variety of conditions and substrates that may be employed in this procedure.

I. Reduction by Lithium–Amine

The reduction of an aromatic system under controlled conditions is an important source of cycloalkanes. The procedure given here employs a solution of lithium in a mixture of low-boiling amines to accomplish that end and affords a mixture of octalins as product. The mixture may be separated by selective hydroboration (Chapter 4, Section III).

OCTALIN FROM NAPHTHALENE (*2*)

A mixture containing 25.6 g (0.2 mole) of naphthalene and 250 ml each of anhydrous ethylamine and dimethylamine is placed in a 1 liter, three-necked, round-bottom flask fitted with a mechanical stirrer and a Dry Ice condenser. After brief stirring, 11.55 g (1.65 g-atom) of lithium wire cut into 0.5-cm pieces is added in one portion. The mixture is stirred for 14 hours, and the Dry Ice condenser is then replaced by a water condenser allowing the solvent to evaporate. Anhydrous conditions are maintained during this process (drying tube attached to the condenser). The flask is then placed in an ice bath and the grayish white residue decomposed by the dropwise addition of about 100 ml of water (*caution!*) accompanied by occasional slow stirring. The mixture is filtered by vacuum, and the residue washed four times with 30-ml portions of ether. The ether layer is separated and the aqueous layer extracted several more times with 25-ml portions of ether. The combined ether extracts are dried over anhydrous calcium sulfate, the solvent is removed, and the residual liquid is distilled. The product (about 20 g, 80%) is collected at 72–77°/14 mm or 194–196°/1 atm and contains 80% $\Delta^{9,10}$-octalin and 20% $\Delta^{1(9)}$-octalin by glpc analysis. The pure $\Delta^{9,10}$-isomer may be obtained by selective hydroboration (Chapter 4, Section III).

II. Reduction by Lithium–Ethylenediamine

A modification of the preceding preparation employs ethylenediamine as a convenient nonvolatile solvent and Tetralin as the commercially available starting material. The results of the reduction are essentially identical.

OCTALIN FROM TETRALIN (*3*)

In a three-necked round-bottom flask fitted with a glass stopper, a stirrer, and a reflux condenser (drying tube) are placed 500 ml of ethylenediamine (distilled from sodium hydroxide pellets before use) and 66.1 g (0.5 mole) of Tetralin. Clean lithium wire (21 g, 3 g-atom) is cut into short pieces and a 5-g portion added to the reaction flask. Stirring is begun and in about 20 minutes the lithium begins to dissolve and heat begins to be evolved. When the bulk of the initial lithium charge has dissolved, the remainder of the lithium is added in 3-g portions over a period of about 15 minutes. Near the end of the addition of lithium, the solution develops a blue color. Stirring is continued for an additional 30 minutes, during which time the blue coloration fades to a slate gray color.

The reaction mixture is decomposed by the addition of 200 ml of ethanol over a period of 20 minutes, and the solution is then poured into 2 liters of ice water. The mixture is extracted with several portions of benzene, and the combined benzene extracts are washed with 5% sulfuric acid followed by water. The solvent is removed by distillation at atmospheric pressure, and the product is distilled at atmospheric pressure through a 20 inch column, bp 194–196°, n_D^{25} 1.4950–1.4970, about 48 g (71%). (For isolation of pure $\Delta^{9,10}$-octalin, see Chapter 4, Section III).

III. Reduction of α,β-Unsaturated Ketones by Lithium–Ammonia

In this experiment, advantage is made of the fact that lithium–ammonia reduction usually proceeds to give trans-fused Decalins (4). Thus, hydrogenation of $\Delta^{1(9)}$-octalone-2 over palladium catalyst gives essentially *cis*-2-decalone as the product, whereas the lithium–ammonia reduction of the octalone gives the trans ring fusion.

A. *trans*-2-DECALONE (5)

The following operations should be carried out in a hood.

A 1-liter three-necked flask fitted with a sealed mechanical stirrer, a drying tube filled with soda lime, and an inlet that can be closed to the atmosphere is heated briefly with a luminous flame. Dry nitrogen gas is swept through the system during this heating and for 30 additional minutes. After introducing 20 g (0.133 mole) of $\Delta^{1(9)}$-2-octalone (Chapter 9, Section III) dissolved in 100 ml of commercial anhydrous ether, the anhydrous ammonia source is connected to the flask, and 500 ml of liquid ammonia is run in. The ammonia inlet tube is closed and stirring is cautiously begun. Then, with vigorous stirring, 5 g (0.7 g-atom) of lithium wire cut into small pieces is added over a 3-minute period. After the initial vigorous evolution of ammonia has ceased, the solution is stirred for 10 minutes. Excess ammonium chloride is then added cautiously over a 30-minute period. After the medium turns white and pasty, 100 ml of water is added to dissolve the salts. Heating the resulting mixture for 1 hour on the steam bath serves to expel most of the ammonia. The reaction mixture is extracted six times with 50-ml portions of ether. The combined extracts are washed twice with water, then with

100 ml of 5% hydrochloric acid, again with water, and finally twice with saturated sodium chloride solution. After removal of the ether on the steam bath, the brown residue is taken up in 45 ml of glacial acetic acid. Oxidation of the decalol is effected by slowly adding a solution of 7.8 g of chromic anhydride dissolved in the minimum amount of water. The solution is cooled during the addition so as to maintain the temperature below 30°. After the exothermic reaction has subsided, the oxidation mixture is allowed to stand at room temperature for 2 days. The dark green solution is then heated for 2 hours on the steam bath and, with cooling, mixed with a solution of 15 g of sodium hydroxide in 60 ml of water. The mixture is extracted five times with 30-ml portions of ether, the combined ethereal extracts are washed with water and saturated salt solution, then dried, and the ether is evaporated. The residual ketone is purified by vacuum distillation; the distillate is collected over a 2° range, bp 112–114°/13 mm, n_D^{25} 1.4823–1.4824.

B. *trans*-10-METHYL-2-DECALONE (6)

The procedure given in the preceding experiment can be applied to the reduction of 10-methyl-$\Delta^{1(9)}$-octalone-2 prepared in Chapter 10, Section VI. The product of the reduction has bp 94–96°/3 mm.

IV. Reduction of α,β-Unsaturated Ketones in Hexamethylphosphoric Triamide

Hexamethylphosphoric triamide (HMPT) is a high-boiling solvent particularly satisfactory for dissolving metals or organometallic compounds. It has been found to be an ideal solvent in which to conduct the reduction of α,β-unsaturated ketones by alkali metals.

General Procedure (7)

In a three-necked flask fitted with a thermometer, a stirrer, a condenser (drying tube), and a pressure-equalizing addition funnel, HMPT (25 ml, 0.14 mole) and anhydrous ether (30 ml) are introduced. Finely divided lithium (1.75 g, 0.25 g-atom) is then added in one portion with stirring. After 5 minutes the solution becomes deep blue and the temperature rises. The flask is cooled in an ice bath and maintained at 20–30°.

A solution of 0.1 mole of the α,β-unsaturated ketone dissolved in 15 ml of anhydrous ether is added dropwise, whereupon the solution is rapidly decolorized. The stirring is continued for 4 hours during which time it slowly becomes blue or green.

Excess lithium is destroyed by the *careful* addition of 1–2 ml of ethanol, and hydrolysis of the reaction mixture is then effected by the addition of a mixture of ice (50 g) and water (100 ml). The solution is then acidified to pH 2 by the addition of 5 N hydrochloric acid, followed by rapid stirring for 1 or 2 minutes to hydrolize the HMPT. The aqueous solution is extracted with ether, the ether solution is dried with magnesium sulfate, then filtered, and the ether is evaporated. The product is isolated by distillation of the residue.

Examples

1. 2-Methyl-2-penten-4-one (mesityl oxide): Treatment of 9.8 g (0.1 mole) of the ketone by the above procedure gives on distillation 6.6 g of methyl isobutyl ketone, bp 117–118°/1 atm and a residue of 3 g.

2. Pulegone

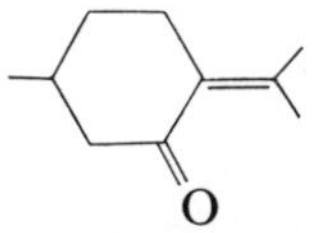

gives 75% of 2-isopropyl-5-methylcyclohexanone, bp 48–53°/0.3 mm, 207°/1 atm.

3. 1-Benzoylcyclohexene gives, after removal of the ether, phenyl cyclohexyl ketone, which is recrystallized from petroleum ether, mp 52° (80%).

4. $\Delta^{1(9)}$-Octalone-2-gives *trans*-2-decalone, bp 112–114°/13 mm.

V. Reduction of an α,β-Unsaturated γ-Diketone with Zinc

The reduction of α,β-unsaturated γ-diketones can conveniently be done with zinc in acetic acid. The following procedure is applicable to the reduction of the Diels-Alder adduct of quinone and butadiene (Chapter 8, Section II).

cis-Δ^2-5,8-OCTALINDIONE (8)

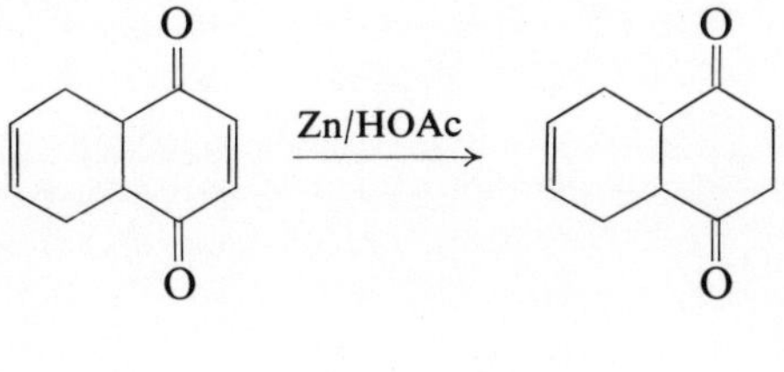

Six grams of the quinone–butadiene adduct are dissolved in 25 ml of 95% acetic acid and the solution is placed in a round-bottom flask fitted with a thermometer and a mechanical stirrer. The flask is immersed in an ice-water bath and rapidly stirred. Small portions of zinc dust (approx. 0.1 g) are added at a rate so as to keep the temperature in the range 30–50°. The addition is discontinued when the temperature ceases to rise as additional quantities of zinc are added (30–40 minutes requiring 2.5–3 g of zinc). Small quantities of acetic acid (3–5 ml) may have to be added to keep the product in solution and the reaction mixture fluid. After completion of the reduction, 20 ml of acetone is added and stirring is continued at room temperature for 5 minutes. The reaction mixture is then filtered under vacuum through celite and the residue washed twice with 10-ml portions of acetone. The filtrate is concentrated under reduced pressure (rotary evaporator or steam bath), and the residue is dissolved in 25 ml of chloroform. The chloroform solution is washed twice with 10-ml portions of water and twice with 10-ml portions of sodium bicarbonate solution, and finally, dried over anhydrous magnesium sulfate and Norit. The solution is filtered and the solvent is evaporated under vacuum. Treatment of the residue with excess ether gives the crystalline product, which is collected and dried in air. The product has mp 100–104°, and the yield is 4.5–5 g (75–83%). The product may be recrystallized from petroleum ether yielding material with a reported (9) mp of 106°.

REFERENCES

1. A. J. Birch, *Quart. Rev.* **4**, 69 (1950); A. J. Birch and H. Smith, *Quart. Rev.* **12**, 17 (1958).
2. R. A. Benkeser and E. M. Kaiser, *J. Org. Chem.* **29**, 955 (1964).
3. L. Reggel, R. A. Friedel, and I. Wender, *J. Org. Chem.* **22**, 891 (1957); W. G. Dauben, E. C. Martin, and G. J. Fonken, *J. Org. Chem.* **23**, 1205 (1958).
4. G. Stork and S. D. Darling, *J. Amer. Chem. Soc.* **86**, 1761 (1964) and references cited therein.
5. E. E. van Tamelen and W. C. Proost, *J. Amer. Chem. Soc.* **76**, 3632 (1954).
6. M. Yanagita, K. Yamakawa, A. Tahara, and H. Ogura, *J. Org. Chem.* **20**, 1767 (1955).
7. P. Angibeaud, M. Larcheveque, H. Normant, and B. Tchoubar, *Bull. Soc. Chim. Fr.*, p. 595 (1968).
8. E. E. van Tamelen, M. Shamma, A. W. Burgstahler, J. Wolinsky, R. Tamm, and P. E. Aldrich, *J. Amer. Chem. Soc.* **91**, 7315 (1969).
9. W. Huckel and W. Kraus, *Chem. Ber.* **92**, 1158 (1959).

4

Hydroboration

A remarkable variation of the hydride reduction is the addition to double bonds of diborane (B_2H_6) (1). Easily generated by the reaction of boron trifluoride etherate with sodium borohydride, the reagent may be used in the generating solution or may be distilled into a receiving flask containing an ether as solvent. Diborane reacts with unsaturated polar functional groups with results similar to those of the metal hydride reducing agents. Its most distinctive property, however, is its ability to add to isolated carbon–carbon double bonds to form alkyl boranes, which may be hydrolized to hydrocarbons or oxidized to alcohols or carbonyl compound according to the reactions. (Alkyl boranes also react with a variety of other reagents in carbon–carbon bond forming reactions. Chapter 12 provides several illustrations).

The procedures given below are typical of the variety of applications of the hydro-boration reaction.

I. Hydroboration of Olefins as a Route to Alcohols

A. CYCLOHEXANOL FROM CYCLOHEXENE: *In Situ* GENERATION OF DIBORANE IN DIGLYME (2)

A 250-ml three-necked flask is fitted with a dropping funnel, a condenser, and a magnetic stirrer. In the flask is placed a solution of cyclohexene (6.5 g, 8.0 ml) in diglyme* (15 ml) and to this is added a solution of sodium borohydride (1.0 g) in diglyme (25 ml). Stirring is begun, and a solution of freshly distilled boron trifluoride diethyl etherate (4.5 g, 4.6 ml) in diglyme (10 ml) is added from the dropping funnel over a period of about 15 minutes. The reaction mixture is now stirred for an additional 20 minutes, and water (10 ml) is carefully added to destroy the slight excess of boro-hydride. When no further hydrogen is evolved, the solution is made alkaline by the addition of 15 ml of dilute sodium hydroxide solution, followed by 15 ml of 30% hydrogen peroxide solution added in 2–3 ml portions.

The reaction mixture is now poured into a separatory funnel with 50 ml of ice water, and the cyclohexanol is removed by two ether extractions. The ether extracts are dried with anhydrous sodium sulfate, and the dried solution is distilled. Cyclohexanol is collected at 154–160°/750 mm, expected yield 5–6 g.

B. CYCLOHEXYLCARBINOL FROM METHYLENECYCLOHEXANE

The preceding method may be applied to an equivalent amount of methylene cyclohexane (7.7 g) with analogous results. The product, cyclohexylcarbinol, has bp 91–92°/23 mm.

C. 4-METHYL-1-PENTANOL FROM 4-METHYL-1-PENTENE: *In Situ* GENERATION OF DIBORANE IN TETRAHYDROFURAN (3)

$$(CH_3)_2CHCH_2CH{=}CH_2 \xrightarrow{B_2H_6}$$

$$(C_5H_{11}{-}CH_2)_3B \xrightarrow[NaOH]{H_2O_2} (CH_3)_2CHCH_2CH_2CH_2{-}OH$$

* Diethylene glycol dimethyl ether, bp 162°.

A 500-ml three-necked flask is fitted with a dropping funnel, a condenser, and a magnetic stirrer. The flask is charged with a mixture of 3.4 g (0.09 mole) of powdered sodium borohydride, 150 ml of THF, and 25.2 g (0.30 mole) of 4-methyl-1-pentene. A solution of 15.1 ml (17.0 g, 0.12 mole) of boron trifluoride etherate in 20 ml of THF is added over a period of 1 hour, the temperature being maintained at 25° (water bath). The flask is stirred an additional hour at 25° and the excess hydride is decomposed with water (10 ml).

The trialkylborane is oxidized by the addition to the stirred reaction mixture of 32 ml of a 3 N solution of sodium hydroxide, followed by the dropwise addition of 32 ml of 30% hydrogen peroxide at a temperature of 30–32° (water bath). The reaction mixture is saturated with sodium chloride and the tetrahydrofuran layer formed is separated and washed with saturated sodium chloride solution. The organic solution is dried over anhydrous magnesium sulfate and the THF is removed. Distillation affords 24.5 g (80%) of 4-methyl-1-pentanol, bp 151–153°/735 mm.

D. *exo*-Norborneol from Norbornene: External Generation of Diborane (3)

A 500-ml three-necked flask is fitted with a thermometer, a condenser, and a gas dispersion tube. A tube from the condenser outlet dips below the surface of mercury in a side-arm test tube. The mercury is covered with a layer of acetone, which serves to destroy excess diborane by the reaction to form diisopropoxyborane, $[(CH_3)_2CHO—]_2BH$ (Fig. 4.1).

A solution of 28.2 g (0.30 mole) of norbornene in 100 ml of THF is placed in the flask. The gas dispersion tube is immersed in the reaction liquid and connected with Tygon tubing to a 250-ml three-necked flask serving as the diborane generator.

The generator is fitted with a pressure-equalizing dropping funnel (acting also as a nitrogen inlet) charged with a solution of sodium borohydride (3.4 g, 20% excess) in 90 ml of diglyme. In the generator is placed a mixture of 23 ml of boron trifluoride etherate (25.5 g, 0.18 mole, 50% excess) and 20 ml of diglyme.

The sodium borohydride solution is added dropwise to the stirred boron trifluoride etherate–diglyme solution resulting in the formation of diborane. The gas is swept into the olefin–THF solution (held at 20°) by maintaining a slow flow of dry nitrogen through the generator.

After completion of the sodium borohydride addition (1 hour), the generator is heated for 1 hour at 70–80° with the nitrogen flow continued to ensure the complete transfer of the diborane. After cooling, the generator is disconnected from the hydro-boration flask.

The excess diborane in the hydroboration flask is decomposed by the cautious addition of 20 ml of water. The organoborane is oxidized by the addition of 32 ml of 3 N sodium hydroxide, followed by dropwise addition of 32 ml of 30% hydrogen peroxide

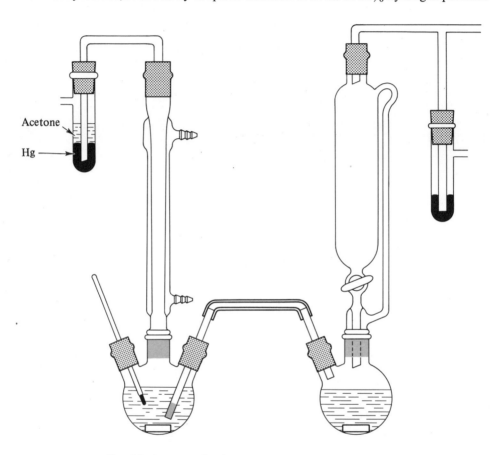

Acetone

Hg

Fig. 4.1. Apparatus for the external generation of diborane.

to the stirred solution maintained at 30–50° (water bath). The reaction mixture is stirred for an additional hour, and 100 ml of ether is added. The ether phase is separated, and the aqueous phase is saturated with sodium chloride and extracted twice with 50-ml portions of ether. The combined ether extracts are washed twice with 50-ml portions of saturated sodium chloride solution and dried over anhydrous magnesium sulfate. The solvent is removed (rotary evaporator), and the residue is crystallized from petroleum ether. The melting point of the sublimed material is 126–127°.

E. (+)-2-BUTANOL FROM *cis*-2-BUTENE: THE USE OF DIISOPINOCAMPHEYLBORANE TO INDUCE ASYMMETRY (*3*)

1. A 500-ml flask is fitted with a condenser, a pressure-equalizing dropping funnel, a magnetic stirrer, and a thermometer and charged with a solution of sodium borohydride (2.85 g) in 75 ml of diglyme. (−)-α-Pinene (27.2 g, 0.2 mole) in 100 ml of diglyme is added in one portion and the flask is cooled in an ice bath with stirring. Boron trifluoride etherate (12.6 ml, 14.2 g, 0.10 mole) is added dropwise to the mixture, and the stirring is continued for an additional 4 hours at 0–5°. Diisopinocampheylborane separates as a thick white precipitate.

2. *cis*-2-Butene (8.5 ml, 6.1 g, 0.11 mole) is condensed in a trap cooled in Dry Ice. Gradual warming allows it to distil into the stirred diisopinocampheylborane, the reaction flask being fitted with a Dry Ice condenser in order to minimize the loss of *cis*-2-butene. The reaction mixture is stirred for 2 hours at ice-bath temperature, then allowed to warm gradually to room temperature over a period of 2 hours. Finally, water (10 ml) is added to decompose the excess hydride.

The alkylborane is oxidized by the addition of 32 ml of 3 N sodium hydroxide followed by 32 ml of 30% hydrogen peroxide to the stirred solution maintained at 30–50° (water bath), and the stirring is continued at the temperature for 1 hour.

The mixture is extracted three times with 100-ml portions of ether, and the combined extracts are washed with saturated aqueous sodium chloride and dried over anhydrous magnesium sulfate. The ether is evaporated (rotary evaporator) and the residue distilled giving 6.15 g (83%) of 2-butanol, bp 98°/725 mm, n_D^{20} 1.3970, $[\alpha]_D^{20}$ + 11.6°.

II. Selective Hydroborations Using Bis(3-methyl-2-butyl)borane (BMB)

Diborane itself is a relatively selective reagent giving largely anti-Markovnikov addition to olefins. 1-Hexene, for example, yields 94% 1-hexanol and only 6% 2-hexanol when treated by diborane followed by peroxide oxidation. The selectivity can be even further increased by the use of certain dialkyl boranes to carry out the hydroboration procedure. Treatment of 1-hexene with BMB, for example, gives after oxidation, 99% 1-hexanol and 1% 2-hexanol. Moreover, BMB is found to react only very slowly with di- or trisubstituted olefins, and can therefore be conveniently used as a selective hydroborating agent in the case of variously substituted polyenes. The generation and use of the reagent are described in the procedures.

A. *n*-OCTANAL FROM 1-OCTYNE (*3*)

$$(CH_3)_2C\!=\!CHCH_3 \xrightarrow{\;B_2H_6\;} (iso\text{-}C_5H_{11})_2BH$$

$$C_6H_{13}C\!\equiv\!CH + (iso\text{-}C_5H_{11})_2BH \;\rightarrow\; C_6H_{13}\!-\!CH\!=\!CH\!-\!B(iso\text{-}C_5H_{11})_2$$

$$\xrightarrow[\text{NaOH}]{H_2O_2} \; n\text{-}C_7H_{15}\!-\!CHO + iso\text{-}C_5H_{11}\!-\!OH$$

A 500-ml three-necked flask is fitted with a condenser, a pressure-equalizing dropping funnel, a magnetic stirrer, and a thermometer. The flask is charged with a mixture of 33.6 g (0.48 mole) of 2-methyl-2-butene and 180 ml of a 1 *M* solution of sodium boro-hydride in diglyme. The flask is cooled in an ice bath and stirring begun. Boron trifluoride etherate (0.24 mole) is added dropwise to the mixture and the solution is stirred at 0° for 2 hours.

The flask is now cooled in an ice–salt bath and 22 g (0.20 mole) of 1-octyne in 20 ml of diglyme is added rapidly (but the temperature is kept at 0–10°). Gradual warming to room temperature completes the hydroboration.

The alkylborane is then oxidized by the addition of 150 ml of a 15% solution of hydrogen peroxide, while the pH of the reaction mixture is maintained at 7–8 by the simultaneous addition of 3 *N* sodium hydroxide, the process being carried out at ice-bath temperature. The reaction mixture is neutralized and subjected to steam distillation. The distillate is extracted with ether, and the extract is dried over anhydrous magnesium sulfate. After removal of the ether, distillation yields 18.0 g (70%) of *n*-octanal, bp 83–85°/33 mm.

B. SELECTIVE HYDROBORATION OF *d*-LIMONENE (*4*)

To 0.165 mole of BMB (prepared as in the preceding experiment) maintained at 0°, is added 20.4 g (0.15 mole) of *d*-limonene over a period of 5 minutes. The reaction mixture is allowed to stand at room temperature for approximately 3 hours. It is then oxidized by the addition of 50 ml of 3 *N* sodium hydroxide followed by 50 ml of 30% hydrogen peroxide. The alcohol is worked up in the usual manner. Upon distillation, the primary "terpineol" is obtained, bp 115–116°/10 mm.

C. Selective Hydroboration of 4-Vinylcyclohexene (4)

In a three-necked flask is placed a solution of 16.2 g of 4-vinyl-cyclohexene (0.15 mole) in 30 ml of diglyme. The flask is cooled with stirring in an ice bath. In a second flask is prepared 0.165 mole of BMB as described above. This second flask is connected to the olefin-containing flask by a short length of glass tubing. By application of a slight pressure of nitrogen, the BMB solution is passed into the cooled olefin solution over a period of approximately 1 hour. After standing an additional hour at room temperature, 10 ml of water is added to destroy excess hydride. Oxidation is carried out by adding 50 ml of 3 N sodium hydroxide followed by 50 ml of 30% hydrogen peroxide, the temperature during the oxidation being kept below 50°. The reaction mixture is worked up in the usual way and affords on distillation 2-(4-cyclohexenyl)-ethanol, bp 86–87°/6 mm.

III. Purification of a Mixture of $\Delta^{9,10}$- and $\Delta^{1(9)}$-Octalins

In experiments given elsewhere (see Chapter 3, Section I or II, or Chapter 7, Section III), a mixture of octalins is prepared by the reduction of naphthalene or Tetralin or by the dehydration of 2-decalol. The following procedure is a convenient method of separating the isomers using BMB as a selective hydroborating agent. The tetra-substituted olefin ($\Delta^{9,10}$) fails to react with BMB under the conditions of the reaction while the trisubstituted olefin ($\Delta^{1(9)}$) forms an adduct with BMB. Oxidation of the resulting alkyl borane gives an alcohol, which can readily be separated from the unreacted $\Delta^{9,10}$-octalin.

$\Delta^{9,10}$-Octalin from a Mixture of Octalins (5)

In a 1-liter, three-necked, round-bottom flask equipped with a magnetic stirrer, dropping funnel, and a reflux condenser attached to a mercury trap is placed a mixture of 4.7 g (0.125 mole) of sodium borohydride, 23.1 g (0.33 mole) of 2-methyl-2-butene, and 100 ml of anhydrous tetrahydrofuran. This mixture is stirred for 15 minutes and then 23.5 g (0.165 mole) of boron trifluoride etherate dissolved in 22 ml of anhydrous tetrahydrofuran is added dropwise over a 45 minute period. The octalin mixture (20 g, 0.15 mole) is then added dropwise over a 10 minute period. After the mixture is stirred for $3\frac{1}{2}$ hours, 50 ml of water is added dropwise with slow stirring. Thirty-five milliliters of 3 N sodium hydroxide is next added dropwise over a 10 minute period, followed by 35 ml of 30% hydrogen peroxide over a 45 minute period. After stirring for 5 hours at 40–45°, the mixture is cooled and the layers separated. The ether layer is washed several times with 30-ml portions of water and dried over anhydrous magnesium sulfate. The ether is removed, and the residual liquid is distilled. The fraction boiling at 75–77°/14 mm is collected (10–13 g).

REFERENCES

1. H. C. Brown, "Hydroboration." Benjamin, New York, 1962.
2. W. Kemp, "Practical Organic Chemistry," p. 112. McGraw-Hill, New York, 1967.
3. G. Zweifel and H. C. Brown, *Org. React.* **13**, 1 (1963).
4. H. C. Brown and G. Zweifel, *J. Amer. Chem. Soc.* **83**, 1241 (1961).
5. R. A. Benkeser and E. M. Kaiser, *J. Org. Chem.* **29**, 955 (1964).

5

Catalytic Hydrogenation

The technique of catalytic hydrogenation can be applied almost universally to unsaturated systems, and therein lies its chief advantage (*1*). By appropriate selection of catalyst, pressure, and temperature, a remarkable variety of substrates can be made to undergo hydrogenation, many of them under hydrogen pressure not exceeding 50 psi (see Appendix 3 for description and use of low-pressure hydrogenation apparatus).

A recent development in the technique of hydrogenation has been the use of homogeneous catalysts. The catalysts employed are soluble in organic solvents and allow for more rapid reactions under milder conditions. The procedures given are typical of low-pressure reactions.

I. Hydrogenation over Platinum Catalyst (*2a*)

Dicyclopentadiene can be hydrogenated conveniently over a platinum catalyst in a Parr apparatus. The tetrahydro product is used in the synthesis of adamantane (Chapter 15, Section I).

endo-TETRAHYDRODICYCLOPENTADIENE FROM THE DIENE (*2b*)

Technical grade dicyclopentadiene is purified by distillation under aspirator pressure through a short column. The fraction boiling at 64–65°/14 mm or 72–73°/22 mm is used in the reduction.

Dicyclopentadiene (50 g, 0.38 mole) is dissolved in 100 ml of anhydrous ether. Platinum oxide (0.25 g) is added, and the mixture is hydrogenated in a Parr apparatus at an initial pressure of 50 psi. Initially the reaction mixture becomes warm. The absorption of 2 mole equivalents of hydrogen takes 4–6 hours. The mixture is filtered by suction to remove the catalyst, and the filtrate is distilled at atmospheric pressure through a short fractionating column.

When the removal of the ether is complete, the condenser is drained of water and the receiving flask is immersed in an ice bath. The condenser is heated by steam, if necessary, to prevent solidification of the distillate in the condenser. The distillation is resumed affording *endo*-tetrahydrodicyclopentadiene, bp 191–193°, about 50 g (98%). The melting point varies with the purity of the starting diene but is usually above 65°.

II. Low-Pressure Hydrogenation of Phenols over Rhodium Catalysts

The hydrogenation of phenols appears to be a straightforward route to cyclohexanols, but the necessary conditions are frequently inconvenient, requiring high pressures (e.g., 100 atm) and high temperatures. However, the development of rhodium catalysts has alleviated these difficulties, allowing the low-pressure (50 psi) hydrogenation of phenols without attendant hydrogenolysis. The catalysts are extremely active, and exposure to explosive mixtures (e.g., hydrogen–oxygen or methanol–oxygen) is to be avoided. Mild evacuation of the reaction vessel containing the catalyst prior to the introduction of hydrogen, and the flushing and evacuation of the vessel several times with hydrogen at low pressure will serve to prevent any difficulties.

A. 1,3-Cyclohexanedione from Resorcinol (3)

$$\text{resorcinol} \xrightarrow[\text{H}_2,\ \text{NaOH}]{\text{Rh/Al}_2\text{O}_3} \text{1,3-cyclohexanedione}$$

A solution of resorcinol (11 g) in sodium hydroxide solution (4.8 g of sodium hydroxide in 20 ml of water) is hydrogenated in the presence of 1.1 g of 5% rhodium on alumina for 16–18 hours at 50 psi initial pressure in a Parr apparatus. The reduction stops after the absorption of 1 equivalent of hydrogen. The catalyst is removed by filtration through celite, and the aqueous solution is carefully acidified with concentrated hydrochloric acid at 0°. The crude product is collected by filtration, dried in air, and recrystallized from benzene to give 1,3-cyclohexanedione, mp 105–107°.

B. *cis*- AND *trans*-1,4-Cyclohexanediol from Hydroquinone (3)

$$\text{HO-}\bigcirc\text{-OH} \xrightarrow[\text{H}_2,\ \text{HOAc}]{\text{Rh/Al}_2\text{O}_3} \text{HO-}\bigcirc\text{-OH}$$

A solution of hydroquinone (11.0 g) in 50 ml of acetic acid containing 1.1 g of 5% rhodium on alumina is hydrogenated at 50 psi in a Parr apparatus. Three equivalents of hydrogen are rapidly absorbed. The catalyst is removed by filtration through celite, and the solvent is evaporated. The oily residue is dissolved in a minimum amount of petroleum ether, and crystallization is induced. The mixture of *cis*- and *trans*-cyclohexanediol has mp 98–100°.

III. *cis*-4-Hydroxycyclohexanecarboxylic Acid from *p*-Hydroxybenzoic Acid

The rhodium catalyst previously discussed is employed in the hydrogenation of *p*-hydroxybenzoic acid. The resulting mixture of cis and trans products is separated by virtue of the ready formation of the lactone of the cis product, which is then hydrolized to the hydroxy acid.

cis-4-HYDROXYCYCLOHEXANECARBOXYLIC ACID (*4*)

p-Hydroxybenzoic acid (55.2 g, 0.4 mole) is dissolved in 175 ml acetic acid and placed in the Parr apparatus. Rhodium on alumina (4 g of 5% catalyst) is added, and the system is briefly evacuated. Hydrogen is introduced to a pressure of 50 psi and shaking is commenced. The absorption of 3 equivalents of hydrogen requires up to 3 days. At the conclusion of the hydrogenation, the catalyst is removed by filtration through celite, and the solution is distilled at atmospheric pressure through a short Vigreux column. The fraction collected boils at 220–260°. It may be partially liquid, but it may be recrystallized from petroleum ether affording *cis*-4-hydroxycyclohexanecarboxylic acid lactone, mp 125–126°. However, the recrystallization is not necessary for conversion to the hydroxy acid.

The lactone (fraction boiling at 220–260°) is dissolved in a minimum amount of water and heated on a steam bath for 2 hours. The cooled solution is extracted six times with small portions of ether (or, preferably, continuously extracted with ether overnight). The ether extracts are dried (anhydrous sodium sulfate), and the ether is evaporated. The residue is recrystallized from acetonitrile giving *cis*-4-hydroxycyclohexanecarboxylic acid, mp 150–152°.

IV. 3-Isoquinuclidone from *p*-Aminobenzoic Acid

Closely related to the use of rhodium catalysts for the hydrogenation of phenols is their use in the reduction of anilines. The procedure gives details for the preparation of the catalyst and its use to carry out the low-pressure reduction of *p*-aminobenzoic acid. Then, as in the preceding experiment, advantage is taken of the formation of a cyclic product to carry out the separation of a mixture of cis and trans cyclohexyl isomers.

3-ISOQUINUCLIDONE (5)

1. *Catalyst:* A mixture of 5.26 g of rhodium chloride trihydrate, 0.34 g of palladium chloride, 18 g of carbon (Darco G-60), and 200 ml of water is rapidly stirred and heated to 80°. A solution of lithium hydroxide hydrate (2.7 g) in 10 ml of water is added in one portion and the heating discontinued. Stirring is continued overnight, after which the mixture is filtered and washed with 100 ml of 0.5% aqueous acetic acid. The product is dried in a vacuum oven at 65°. About 20 g of the catalyst is thus obtained.

2. *Hydrogenation:* A mixture of 27.4 g (0.20 mole) of *p*-aminobenzoic acid, 200 ml of water, and 2 g of the catalyst is hydrogenated at 50 psi in a Parr apparatus. At the completion* of the hydrogenation (absorption of 0.6 mole of hydrogen), the mixture is filtered and concentrated under vacuum. When crystals start to form, the mixture is diluted with 200 ml of DMF and cooled in an ice bath. The crystals are collected by filtration, washed with DMF followed by methanol, and dried. About 20 g of *cis*- and *trans*-4-aminocyclohexanecarboxylic acid, mp 292–296°, are obtained.

3. *3-Isoquinuclidone:* A 5-g portion of *cis*- and *trans*-4-aminocyclohexanecarboxylic acid is mixed with 30 ml of Dowtherm A† and heated rapidly to reflux in a flask fitted with a short distilling column. Water distils during the heating, which is continued for about 20 minutes. At this time, the mixture is homogeneous. The cooled solution is

---

* The rate of reaction is variable requiring from 1–4 days. Fresh catalyst is added whenever the rate of hydrogen uptake markedly decreases. Added catalyst must first be wet with solvent. The hydrogen must be well evacuated, for opening the mixture to the atmosphere without such evacuation will produce a mixture that may explode on contact with fresh catalyst.

† A eutectic mixture of diphenyl and diphenyl ether, available from Dow Chemical Co.

diluted with 100 ml of isooctane and extracted three times with 50-ml portions of water. The combined aqueous extracts are clarified with charcoal, filtered, and evaporated to dryness under vacuum. The residue, after recrystallization from cyclohexane, affords about 3.5 g of 3-isoquinuclidone, mp 197–198°.

V. Homogeneous Catalytic Hydrogenation

In benzene or similar solvents, tris(triphenylphosphine)halogenorhodium(I) complexes, $RhX[P(C_6H_5)_3]_3$, are extremely efficient catalysts for the homogeneous hydrogenation of nonconjugated olefins and acetylenes at ambient temperature and pressures of 1 atmosphere (6). Functional groups (keto-, nitro-, ester, and so on) are not reduced under these conditions.

Some generalizations that pertain are: (1) Terminal olefins are more rapidly reduced than internal olefins; (2) conjugated olefins are not reduced at 1 atmosphere; (3) ethylene is not hydrogenated. Rates of reduction compare favorably with those obtained by heterogeneous catalysts such as Raney nickel or platinim oxide. In fact, the hydrogenation of some olefins may be so rapid that the temperature of the solution (benzene) is raised to the boiling point.

One useful feature of this reducing system is its apparent ability to allow deuteration of double bonds without "scrambling." Although the precise stereochemistry of the addition remains to be established, the incorporation of only two deuterium atoms per double bond has been clearly demonstrated (7).

The preparation of one of the complexes is given, as well as some examples of its use.

A. TRIS(TRIPHENYLPHOSPHINE)CHLORORHODIUM(I) (8)

$$RhCl_3 + 4\ P(C_6H_5)_3 \rightarrow RhCl[P(C_6H_5)_3]_3 + Cl_2P(C_6H_5)_3$$

$$Cl_2P(C_6H_5)_3 + H_2O \rightarrow OP(C_6H_5)_3 + 2\ HCl$$

A 500-ml round-bottom flask is equipped with a reflux condenser, a gas inlet tube, and a gas outlet leading to a bubbler. The flask is charged with a solution of rhodium (III) chloride trihydrate (2 g) in 70 ml of 95% ethanol. A solution of triphenylphosphine (12 g, freshly recrystallized from ethanol to remove the oxide) in 350 ml of hot ethanol is added to the flask, and the system is flushed with nitrogen. The mixture is refluxed for 2 hours, following which the hot solution is filtered by suction to obtain the product. The crystalline residue is washed with several small portions of anhydrous ether (50 ml total) affording the deep red crystalline product in about 85% yield.

B. HYDROGENATIONS EMPLOYING TRIS(TRIPHENYLPHOSPHINE)CHLORORHODIUM(I) (TTC) (7, 9)

$$RCH\!=\!CHR' \xrightarrow[H_2]{TTC} RCH_2\!-\!CH_2R'$$

General Procedure

A 500-ml round-bottom flask is fitted with a magnetic stirrer and a three-way stopcock attached to a hydrogen source and a vacuum source. The flask is charged with a solution of the olefin (0.005 mole) in dry benzene (20–40 ml), and 0.1 g of the catalyst is added. The flask is evacuated and then filled with hydrogen at atmospheric pressure a total of three times. Stirring is then begun and continued until hydrogenation is complete (from 1–20 hours). Stirring is then discontinued, the flask is vented to the air and finally flushed briefly with air. The solvent is evaporated (rotary evaporator) and the residue is taken up in methylene chloride–petroleum ether (1 : 1) and filtered through alumina. Evaporation of this solvent mixture affords the product.

Examples

1. Cyclohexene is hydrogenated to cyclohexane under the above conditions in about 15 minutes.

2. Ergosterol (a triene) rapidly takes up 1 mole of hydrogen selectively to give 5α,6-dihydroergosterol, recrystallized from methanol–chloroform, mp 173–174°.

3. Geranyl acetate (a diene) takes up 2 moles of hydrogen unselectively in 48 hours to give the saturated acetate, 3,7-dimethyloctyl acetate, bp 109–110°/12 mm, n_D^{20} 1.4261. (Geraniol itself has an allylic hydroxyl and appears to suffer decarbonylation under these reaction conditions.)

4. β-Nitrostyrene requires from 60–90 hours for the uptake of 1 mole of hydrogen. The product, 1-phenyl-2-nitroethane, is purified by distillation, bp 76–77°/0.5 mm.

5. Methyl oleate gives methyl 9,10-dideuterostearate on treatment with deuterium. Similarly, methyl linoleate gives 9,10,12,13-tetradeuterostearate. The products have bp 214–215°/15 mm.

6. 1,4-Naphthoquinone reacts in 30 minutes to give 1,2,3,4-tetrahydro-1,4-dioxo-naphthalene in 70% yield. After recrystallization from hexane, the product has mp 95–97° (*10*).

REFERENCES

1. F. J. McQuillin, *in* A. Weissberger, ed., "Technique of Organic Chemistry," Vol. II. Wiley/Interscience, New York, 1963.
2a. P. N. Rylander, "Catalytic Hydrogenation over Platinum Metals." Academic Press, New York, 1967.
2b. P. von R. Schleyer, M. M. Donaldson, R. D. Nicholas, and C. Cupas, *Org. Syn.* **42**, 8 (1962).
3. J. C. Sircar and A. I. Meyers, *J. Org. Chem.* **30**, 3206 (1965).
4. D. S. Noyce, G. L. Woo, and B. R. Thomas, *J. Org. Chem.* **25**, 260 (1960).
5. W. M. Pearlman, *Org. Syn.* **49**, 75 (1969).
6. J. A. Osborn, F. H. Jardine, J. F. Young, and G. Wilkinson, *J. Chem. Soc.* (*A*), p. 1711 (1966).
7. A. J. Birch and K. A. M. Walker, *J. Chem. Soc.* (*C*), p. 1894 (1966).
8. J. A. Osborn and G. Wilkinson, *Inorg. Syn.* **10**, 67 (1967).
9. C. Djerassi and J. Gutzwiller, *J. Amer. Chem. Soc.* **88**, 4537 (1966).
10. A. J. Birch and K. A. M. Walker, *Tetrahedron Lett.*, p. 3457 (1967).

6

The Introduction of Halogen

The introduction of halogen into organic molecules can be carried out by a variety of addition or substitution reactions. The classical methods for the addition of halogen to double bonds or the substitution of halogen for hydroxyl by hydrohalic acids are too well known to bear repetition here. Discussed below, then, are methods that are of interest because of their stereospecific outcome or because they may be used on sensitive substrates.

I. Halides from Alcohols by Triphenylphosphine–Carbon Tetrahalide

Triphenylphosphine in refluxing carbon tetrachloride converts primary and secondary alcohols into the corresponding chlorides under very mild conditions (1) with inversion of configuration (2, 3). A suggested route for the transformation is shown.

$$\phi_3P + CCl_4 \rightarrow \phi_3P^{\oplus}-CCl_3(Cl^{\ominus}) \xrightarrow[(-CHCl_3)]{R-OH}$$

$$\phi_3P^{\oplus}-O-R(Cl^{\ominus}) \rightarrow R-Cl + \phi_3PO$$

When carbon tetrabromide is used, the alkyl bromide is formed. Providing moisture is excluded from the reaction mixture (HX is formed otherwise), the reaction conditions are completely neutral, affording a convenient preparation of the halides of acid-sensitive substrates (for example, sugars).

General Procedure and Examples (2, 3)

$$R-OH + CCl_4 + (C_6H_5)_3P \rightarrow R-Cl + CHCl_3 + (C_6H_5)_3PO$$

Carbon tetrachloride is dried over calcium chloride and distilled before use. Triphenylphosphine is dried by dissolving it in dry benzene, removing the solvent under vacuum, and storing over phosphorus pentoxide. Alcohols should also be dried before use.

In a well-dried, 500-ml, round-bottom flask are placed the alcohol (0.1 mole), triphenylphosphine (29 g, 0.11 mole), and 200 ml of anhydrous carbon tetrachloride. The flask is fitted with a condenser (drying tube) and refluxed (steam bath) for 2–3 hours. The flask is then cooled to room temperature and 200 ml of pentane is added to precipitate the bulk of the triphenylphosphine oxide. The mixture is filtered and the filtrate is fractionally distilled affording the product. Examples are given in Table 6.1 (*2, 3*).

TABLE 6.1

Alcohol	bp/1 atm of chloride (°C)	Yield (%)
1-Butanol	78	89
2-Butanol	68	65
Neopentyl alcohol	85	76
1-Decanol	223	78
Benzyl alcohol	178	83

II. Halides from Alcohols and Phenols by Triphenylphosphine Dihalide

A halogenating system related to the preceding case is formed by the reaction of triphenylphosphine with molecular bromine or chlorine. The system is not as sensitive to moisture as the phosphine–carbon tetrahalide system (see preceding section), but it suffers from the disadvantage that hydrohalic acids are produced as the reaction proceeds. Nevertheless, sensitive compounds can be successfully halogenated by the system, as exemplified by the preparation of cinnamyl bromide from the alcohol.

An important feature of this reagent is its ability to convert phenols to the corresponding bromides under laboratory conditions. Examples of these conversions are given.

A. *n*-Butyl Bromide and Other Examples (*4*)

$$n\text{-}C_4H_9\text{—}OH + (C_6H_5)_3PBr_2 \xrightarrow{\text{DMF}} n\text{-}C_4H_9\text{—}Br + (C_6H_5)_3PO + HBr$$

A 500-ml, three-necked, round-bottom flask is fitted with a mechanical stirrer, a pressure-equalizing addition funnel and a reflux condenser (drying tube). The flask is charged with 100 ml of dry dimethylformamide, 9.2 ml (0.1 mole) of dry *n*-butyl alcohol, and 28 g (0.107 mole) of dry triphenylphosphine. The flask is cooled in a water bath, stirring is begun, and bromine (approx. 16 g, 0.1 mole) is added over about 15 minutes.

During the addition, the flask temperature is held below 55° by addition of ice to the cooling bath as necessary. The addition of bromine is halted when two drops give a persistent orange color.

Stirring and cooling are stopped and the flask is set up for vacuum distillation. All volatile materials are then distilled under a vacuum of 5 mm into a Dry Ice cooled receiver. The distillate is diluted with 500 ml of water, and the organic phase is separated. The crude product is dried over anhydrous magnesium sulfate and distilled affording pure *n*-butyl bromide, bp 100–102°. (The yield is about 90% prior to the final distillation.)

The procedure for the preparation of other alkyl bromides is the same. The following alkyl bromides have been prepared by this reaction in the yields shown in Table 6.2.

TABLE 6.2

Alcohol	bp/1 atm of bromide (°C)	Yield (%)
Isobutyl	91	89
Neopentyl	109	79
2-Butyl	91	90
Cyclopentyl	137	83
Cyclohexyl	165	88
Cholesterol (5)	(mp 96–98)	80

B. CINNAMYL BROMIDE (6)

$$C_6H_5CH{=}CHCH_2OH + (C_6H_5)_3PBr_2 \xrightarrow{\ CH_3CN\ }$$
$$C_6H_5CH{=}CHCH_2Br + (C_6H_5)_3PO + HBr$$

The setup described in the preceding experiment is employed. The flask is charged with 175 ml of dry acetonitrile and 53.2 g (0.205 mole) of triphenylphosphine. The flask is cooled in an ice-water bath, stirring is begun, and 32 g (0.20 mole) of bromine is added over 15 minutes. (An excess of bromine should not be present in the reaction vessel. If a pink color remains, a small amount of triphenylphosphine is added to consume the excess.) Cooling is discontinued and a solution of cinnamyl alcohol (27 g, 0.20 mole) in 25 ml of acetonitrile is added with stirring over a period of 5 minutes. After the addition is complete, the solution is vacuum distilled to remove solvent. The residue is distilled at 2–4 mm through an air condenser (bp 90–100°). The crude product is taken up in 100 ml of ether, washed with sodium carbonate solution, dried over anhydrous magnesium sulfate, and distilled. The product has bp 66–68°/0.07 mm, mp 29°, and the yield is about 60%.

C. *p*-CHLOROBROMOBENZENE (4)

$$Cl\text{-}\langle\bigcirc\rangle\text{-}OH + (C_6H_5)_3PBr_2 \xrightarrow{200°} Cl\text{-}\langle\bigcirc\rangle\text{-}Br + HBr + (C_6H_5)_3PO$$

Caution: The evolution of HBr in this experiment necessitates the use of a hood or efficient gas trap.

By the procedure described in the preceding experiment, 30 g (0.11 mole) of triphenylphosphine dissolved in 100 ml of acetonitrile is converted to triphenylphosphine dibromide. After the addition of the bromine has been completed, the cooling bath is removed, the flask is set up for vacuum distillation, and the solvent is removed. To the residue is added *p*-chlorophenol (10.3 g, 0.08 mole), and the flask is heated at 200° (mantle, wax bath, or sand bath) until HBr ceases to be evolved (about 2 hours). The flask is cooled and the contents are steam distilled affording crude *p*-chlorobromobenzene in about 90% yield. Recrystallization from benzene gives the pure product, mp 65–66°.

Similarly, this procedure has been applied to phenol (92% yield), *p*-nitrophenol (60% yield), and *p*-methoxyphenol (59% yield).

III. Allylic and Benzylic Bromination with *N*-Bromosuccinimide

As mentioned in an earlier section (cf. Chapter 1, Section III), allylic positions are subject to attack by free radicals resulting in the formation of stable allyl radicals. *N*-Bromosuccinimide (NBS) in the presence of free-radical initiators liberates bromine radicals and initiates a chain reaction bromination sequence by the abstraction of allylic or benzylic hydrogens. Since NBS is also conveniently handled, and since it is unreactive toward a variety of other functional groups, it is usually the reagent of choice for allylic or benzylic brominations (7).

A. 3-BROMOCYCLOHEXENE (8)

$$\bigcirc + \text{ NBS } \xrightarrow{CCl_4} \bigcirc^{Br} + \bigcirc_{O}^{O}\text{NH}$$

A 250-ml round-bottom flask is fitted with a condenser (drying tube), a magnetic stirrer, and a heating mantle. The flask is charged with 8.2 g (0.1 mole) of cyclohexene, 14 g (0.079 mole) of NBS, 0.1 g of benzoyl peroxide, and 50 ml of dry carbon tetrachloride. The flask is flushed with nitrogen and then refluxed for 40 minutes with stirring. The succinimide is removed by suction filtration and washed twice with 10-ml portions of carbon tetrachloride. The combined filtrate and washings are fractionally distilled at atmospheric pressure to remove the carbon tetrachloride and excess olefin (steam bath). The residue is distilled under vacuum, giving about 60% yield of 3-bromo-cyclohexene, bp 68°/15 mm or 44°/2 mm.

B. 4-Bromo-2-heptene (9)

$$CH_3CH_2CH_2CH_2CH{=}CH{-}CH_3 + NBS \xrightarrow{CCl_4}$$

$$CH_3CH_2CH_2\underset{\underset{Br}{|}}{CH}{-}CH{=}CH{-}CH_3 +$$

The above procedure is applied to 2-heptene (9.8 g, 0.1 mole), 11.5 g (0.066 mole) of NBS, and 0.1 g of benzoyl peroxide in 50 ml of carbon tetrachloride. The mixture is refluxed with stirring for 2 hours. Final fractionation yields 50–65% of 4-bromo-2-heptene, bp 70–71°/32 mm.

C. 1-Phenylethyl Bromide (10)

$$C_6H_5CH_2CH_3 + NBS \xrightarrow{CCl_4} C_6H_5\underset{\underset{Br}{|}}{CH}CH_3 +$$

A mixture of ethylbenzene (11.7 g, 0.11 mole), NBS (17.8 g, 0.10 mole), benzoyl peroxide (0.24 g), and carbon tetrachloride (55 ml) is refluxed as in the above experiments for 25 minutes. After removal of the succinimide and solvent, the residue is distilled, giving the product in about 80% yield, bp 87–90°/14 mm.

IV. α-Bromination of Ketones and Dehydrobromination

α-Bromination of ketones proceeds via the acid catalyzed formation of the enol. The enol is attacked by bromine, and finally, a proton is lost giving the product. Acetic acid is a convenient medium for the reaction since it is a proton source as well as an excellent solvent for the reagents. Examples (11) of the bromine–acetic acid system are given below.

$$\underset{\underset{O}{\|}}{R-C-CH_2-R'} \underset{}{\overset{H^+}{\rightleftharpoons}} \underset{OH}{\underset{|}{R-C=CH-R'}} \overset{Br_2}{\longrightarrow}$$

$$\underset{\underset{\overset{\oplus}{O}\diagdown H}{\|}}{R-C-\underset{|}{\overset{Br}{C}}H-R'} + Br^{\ominus} \overset{-H^+}{\longrightarrow} \underset{\underset{O}{\|}}{R-C-\underset{|}{\overset{Br}{C}}H-R'}$$

Base catalyzed dehydrobromination of α-bromoketones is frequently employed in steroidal systems as a means to obtain α,β-unsaturated ketones (12). An example of this reaction employing lithium carbonate as the base is also given.

A. 2-BROMOCHOLESTANONE (13)

$$\xrightarrow[\text{HBr}]{\text{Br}_2/\text{HOAc}}$$

A 100-ml round-bottom flask is equipped with a magnetic stirrer and a pressure-equalizing addition funnel. The flask is charged with a solution of 3.9 g (0.01 mole) of cholestanone (mp 128–130°) in 30 ml of acetic acid containing 1 % (approx. 5 drops) of 48 % aqueous hydrobromic acid. The addition funnel is charged with a solution of 1.7 g (0.011 mole) of bromine in 10 ml of acetic acid, and a few drops of the bromine solution are added to the stirred reaction flask. When the color of the bromine has been discharged, the remainder of the solution is run in dropwise over about a 20-minute period. The solution is then stirred at room temperature for 3 hours. After the stirring period is over, the reaction mixture is poured with stirring into 250 ml of water, and the aqueous solution extracted with 50 ml of ether. The ether extract is washed with sodium bicarbonate solution then dried (anhydrous sodium sulfate), and the ether is evaporated (rotary evaporator). The residue is recrystallized from ethanol giving product of mp 169–170°.

B. trans-3-BROMO-trans-10-METHYL-2-DECALONE (14)

$$\xrightarrow[\text{HBr}]{\text{Br}_2\text{HOAc}}$$

The bromination given in the preceding experiment may be applied to *trans*-10-methyl-2-decalone prepared in Chapter 3, Section III. The product is recrystallized from petroleum ether–benzene (10:1) giving colorless needles, mp 101–103°.

C. DEHYDROBROMINATION OF 2-BROMOCHOLESTANONE: Δ^1-CHOLESTENONE (*15*)

A 250-ml round-bottom flask fitted with a condenser (drying tube) is charged with a mixture of 2-bromocholestanone (4.7 g, 0.01 mole), lithium carbonate (7.4 g, 0.10 mole), and 100 ml of dimethylformamide. The system is flushed with nitrogen and then refluxed (mantle) for 18–24 hours. After the reflux period, the solution is cooled and poured into 500 ml of water. The aqueous mixture is extracted with 50 ml of ether, the ether extract is dried (sodium sulfate), and the ether is removed (rotary evaporator). The residue may be recrystallized from ethanol or methanol. Δ^1-Cholestenone is a white solid, mp 98–100°.

V. Stereospecific Synthesis of *trans*-4-Halocyclohexanols

A mixture of *cis*- and *trans*-1,4-cyclohexanediols obtained from the hydrogenation of hydroquinone can be converted into the bridged 1,4-oxide by dehydration over alumina. The oxide, which is required to have a cis geometry, can then be cleaved by hydrohalic acids to give stereospecifically the trans disubstituted products.

A. 1,4-CYCLOHEXANE OXIDE (*16*)

Neutral alumina (12 g) is placed in a 250-ml flask and partially deactivated by the addition of 0.12 g (2–3 drops) of water. The alumina is allowed to stand with occasional swirling for 30 minutes. To the flask is added 20 g of 1,4-cyclohexanediol (Chapter 5, Section II). A short Vigreux column is placed over the flask, which is then arranged for distillation. The mixture is heated at 240° with a sand bath or mantle, whereupon distillation commences (*caution, bumping*). The distillation is continued until the alumina is dry (5–8 hours). The distillate, which consists of two layers, is added to 75 ml of pentane, and the aqueous layer is drawn off. The pentane solution is dried (anhydrous magnesium sulfate), then filtered, and the solution is fractionated. 1,4-Cyclohexane oxide is collected at 116–119°/1 atm.

B. *trans*-4-CHLOROCYCLOHEXANOL (*17*)

To 10 g of cyclohexane-1,4-oxide is added 40 ml of 12 N hydrochloric acid solution. The solution is mixed thoroughly and allowed to stand at room temperature for 8 days. Water (50 ml) is added to the mixture, the phases are separated, the aqueous phase is extracted with 25 ml of ether, and the ether extract is combined with the organic phase. The ether solution is washed with bicarbonate solution and water and dried over anhydrous sodium sulfate. Ether and unreacted oxide are removed on a rotary evaporator, and the product is recrystallized from petroleum ether, mp 82–83° (yield, 8 g).

C. *trans*-4-BROMOCYCLOHEXANOL (*18*)

To 10 g of cyclohexane-1,4-oxide is added 48% aqueous hydrobromic acid (60 g). The phases are mixed thoroughly and allowed to stand at room temperature until the solution separates into two layers (usually 5 days). The mixture is saturated with sodium chloride and extracted twice with 25-ml portions of ether. The ether layer is washed with an equal volume of saturated sodium bicarbonate solution, then with the same amount of water. Finally, the ether solution is dried over anhydrous sodium sulfate, the ether is evaporated, and the residue is allowed to cool, whereupon crystallization should follow. The crude product may be recrystallized from petroleum ether giving material of mp 81–82° (yield, 11 g).

REFERENCES

1. I. M. Downie, J. B. Holmes, and J. B. Lee, *Chem. Ind. (London)*, p. 900 (1966).
2. J. B. Lee and I. M. Downie, *Tetrahedron* 23, 359 (1967).
3. A. W. Friederang and D. S. Tarbell, *J. Org. Chem.* 33, 3797 (1968).
4. G. A. Wiley, R. L. Hershkowitz, B. M. Rein, and B. C. Chung, *J. Amer. Chem. Soc.* 86, 964 (1964).
5. D. Levy and R. Stevenson, *J. Org. Chem.* 30, 3469 (1965).
6. J. P. Schaefer, J. G. Higgins, and P. K. Shenov, *Org. Syn.* 48, 51 (1968).

7. C. Djerassi, *Chem. Rev.* **43**, 271 (1948); L. Horner and E. H. Winkelmann *in* "Newer Methods of Preparative Organic Chemistry" (W. Foerst, ed.), Vol. 3, Academic Press, New York, 1964.

8. N. S. Isaacs, "Experiments in Physical Organic Chemistry," p. 276. Macmillan, London, 1969.

9. F. L. Greenwood and M. D. Kellert, *J. Amer. Chem. Soc.* **75**, 4842 (1953); F. L. Greenwood, M. D. Kellert, and J. Sedlak, *Org. Syn. Collective Vol.* **4**, 108 (1963).

10. H. J. Dauben and L. L. McCoy, *J. Amer. Chem. Soc.* **81**, 5405 (1959).

11. H. O. House and H. W. Thompson, *J. Org. Chem.* **28**, 360 (1963).

12. H. O. House and R. W. Bashe, *J. Org. Chem.* **30**, 2942 (1965).

13. A. Butenandt and A. Wolff, *Chem. Ber.* **68B**, 2091 (1935).

14. M. Yanagita and K. Yamakawa, *J. Org. Chem.* **21**, 500 (1956).

15. G. F. H. Green and A. G. Long, *J. Chem. Soc.*, p. 2532 (1961).

16. E. A. Fehnel, S. Goodyear, and J. Berkowitz, *J. Amer. Chem. Soc.* **73**, 4978 (1951).

17. E. L. Bennett and C. Niemann, *J. Amer. Chem. Soc.* **74**, 5076 (1952).

18. D. S. Noyce, B. N. Bastian, P. T. S. Lau, R. S. Monson, and B. Weinstein, *J. Org. Chem.* **34**, 1247 (1969).

7

Miscellaneous Elimination, Substitution, and Addition Reactions

The following experimental procedures do not fall into any convenient categories, but all require reagents and techniques of general interest in organic synthesis.

I. Methylenecyclohexane by Pyrolysis of an Amine Oxide

The pyrolysis of amine oxides is a reliable synthesis of unrearranged olefins. Its importance in this regard, however, has largely been reduced by the development of the Wittig reaction (Chapter 11). Nevertheless, the present reaction still plays a role in the degradation of naturally occurring nitrogen-containing compounds, and it is included here as an example of the techniques employed.

METHYLENECYCLOHEXANE (*1*)

A mixture of *N,N*-dimethylcyclohexylmethylamine (49.4 g, 0.35 mole, Chapter 2, Section I), 30% hydrogen peroxide (39.5 g, 0.35 mole) and 45 ml of methanol is placed in a 500-ml Erlenmeyer flask, covered with a watch glass, and allowed to stand at room temperature. After 2 hours, and again after an additional 3 hours, 30% hydrogen peroxide (39.5-g portions each time) is added with swirling. The solution is allowed to stand at room temperature for an additional 30 hours, whereupon excess peroxide is destroyed by the cautious addition (swirling) of a small amount of platinum black. Cessation of oxygen evolution indicates complete decomposition of the excess peroxide.

The mixture is filtered into a 500-ml round-bottom flask, and methanol and water are removed by distillation under vacuum (bath temperature 50–60°) until the residual amine oxide hydrate solidifies. The flask is fitted with a magnetic stirrer and a short Vigreux column, and the receiving flask is cooled in a Dry Ice–acetone bath. The flask is heated (oil bath, contained in stainless steel or copper) to 90–100° and vacuum (approx. 10 mm) is applied with stirring. Resolidification of the amine oxide indicates loss of the water of hydration. Thereupon, the bath temperature is raised to 160° and decomposition of the amine oxide occurs over about 2 hours. The distillate is mixed with water (100 ml) and the olefin layer is separated. The organic phase is washed twice with 5-ml portions of water, twice with 5-ml portions of ice-cold 10% hydrochloric acid, and once with 5 ml of bicarbonate solution. The chilled (Dry Ice–acetone) olefin is filtered through Pyrex wool and distilled from a small piece of sodium. Methylenecyclohexane is collected at 100–102°, yield about 27 g (80%).

II. The Wolff-Kishner Reduction

The method of choice for the reduction of keto groups to methylenes remains the Wolff-Kishner reaction (2) or its modification discovered by Huang-Minlon (3). Originally, the hydrazone formation was carried out in a separate step, followed by

$$R_2C{=}O + H_2N{-}NH_2 \rightarrow R_2C{=}N{-}NH_2 + H_2O$$

$$\xrightarrow[\Delta,\ KOH]{\text{diethylene glycol}} R_2CH_2$$

decomposition with base under pressure. The Huang-Minlon modification allows the reaction to be carried out in a single flask using a high boiling solvent to facilitate the decomposition of the hydrazone.

General Procedure

$$R_2C{=}O \xrightarrow[\text{diethylene glycol}]{H_2N{-}NH_2,\ KOH} R_2CH_2$$

The carbonyl compound to be reduced (0.1 mole) is placed in a 250-ml round-bottom flask with 13.5 g of potassium hydroxide, 10 ml of 85% hydrazine hydrate, and 100 ml of diethylene glycol. A reflux condenser is attached and the mixture is heated to reflux for 1 hour (mantle). After refluxing 1 hour, the condenser is removed and a thermometer is immersed in the reaction mixture while slow boiling is continued to remove water. When the pot temperature has reached 200°, the condenser is replaced and refluxing is continued for an additional 3 hours. The mixture is then cooled, acidified with concentrated hydrochloric acid, and extracted with benzene. The benzene solution is dried, and the benzene is evaporated to afford the crude product, which is purified by recrystallization or distillation.

Examples

1. Benzophenone gives diphenylmethane in 83% yield, bp 149°/29 mm.
2. Propiophenone gives *n*-propylbenzene in 82% yield, bp 160–163°.
3. Cyclohexanone is converted to cyclohexane in 80% yield, bp 80–81°.
4. *trans*-2-Decalone (Chapter 3, Section III) gives *trans*-Decalin, bp 185°, n_D^{20} 1.4685.
5. *trans*-10-Methyl-2-decalone (Chapter 3, Section III) gives *trans*-9-methyldecalin, bp 90–91°/20 mm, n_D^{25} 1.4764 (*4*).

III. Dehydration of 2-Decalol

As previously described, a mixture of $\Delta^{9,10}$- and $\Delta^{1(9)}$-octalins can be prepared by the reduction of naphthalene or Tetralin. Another route to this mixture is the dehydration of a mixture of 2-decalol isomers. This latter route has certain advantages in that one can avoid the handling of lithium metal and low-boiling amines. Moreover, 2-decalol is available commercially or can be prepared by the hydrogenation of 2-naphthol (*5*). In either case a comparable mixture of octalins is obtained, which can be purified by selective hydroboration to give the pure $\Delta^{9,10}$-octalin (Chapter 4, Section III).

OCTALIN FROM 2-DECALOL (*6*)

2-Decalol (25 g of a commercial mixture of isomers) is added to 75 g of 100% phosphoric acid with stirring in a 250-ml, three-necked, round-bottom flask. The flask is arranged for distillation through a short (5–8 inch) fractionating column, and the mixture is heated to 150° with a mantle or oil bath. A slight vacuum is applied to the system from an aspirator, and water is added dropwise through a pressure-equalizing addition funnel. Octalin and water distil and the reaction is continued until no more steam-volatile material is obtained. The distillate is dissolved in ether, the water is separated, and the ethereal layer is dried over anhydrous magnesium sulfate. Evaporation of the solvent on a rotary evaporator followed by distillation at atmospheric pressure affords the octalin mixture, bp 189–193°, yield 75–80%.

IV. Boron Trifluoride Catalyzed Hydrolysis of Nitriles

The hydrolysis of nitriles catalyzed by boron trifluoride is a reliable and high yield process for conversion to the corresponding amide. Other methods give variable yields and may result in a significant quantity of acid being formed, whereas the procedure given below frequently results in yields above 90%.

CONVERSION OF NITRILES TO AMIDES (7)

$$R—CN + H_2O \xrightarrow[BF_3]{HOAc} R—\overset{\displaystyle O}{\underset{\displaystyle \|}{C}}—NH_2$$

A solution of 3 g of the nitrile, water (5 moles per mole of nitrile), and 20 g of boron trifluoride–acetic acid complex is heated (mantle or oil bath) at 115–120° for 10 minutes. The solution is cooled in an ice bath with stirring and is carefully made alkaline by the slow addition of 6 N sodium hydroxide (about 100 ml). The mixture is then extracted three times with 100-ml portions of 1:1 ether–ethyl acetate, the extracts are dried over anhydrous sodium sulfate, and the solvent is evaporated on a rotary evaporator to yield the desired amide. The product may be recrystallized from water or aqueous methanol. Examples are given in Table 7.1.

TABLE 7.1

Nitrile	Amide	mp of amide (°C)
Phenylacetonitrile	Phenylacetamide	154–156
p-Nitrobenzonitrile	p-Nitrobenzamide	198–200
Benzonitrile	Benzamide	126–128
o-Tolunitrile	o-Toluamide	144–145
p-Tolunitrile	p-Toluamide	155–156

V. Bridged Sulfides by Addition of Sulfur Dichloride to Dienes

An interesting addition reaction of sulfur dichloride has been discovered that allows the preparation in good yield of bridged cyclic sulfides from cyclic dienes. Two examples of the reaction have been described (8) employing in one case 1,3-cyclo-octadiene and in the other 1,5-cyclooctadiene. The former sequence is shown in the scheme. The experimental details of the latter sequence are given below.

1,5-CYCLOOCTANE SULFIDE (9-THIABICYCLO[3.3.1]NONANE) FROM 1,5-CYCLOOCTADIENE (8)

1. *2,6-Dichloro-1,5-cyclooctane Sulfide:* A 250-ml, three-necked, round-bottom flask is fitted with two addition funnels and a stirrer, all openings being protected by drying tubes. In one funnel is placed 89 ml (0.72 mole) of 1,5-cyclooctadiene, and in the other is placed 46 ml (0.72 mole assuming 100% purity) of commercial sulfur dichloride. In the flask is placed 90 ml of methylene chloride, which is cooled with stirring to -5 to $0°$ by a Dry Ice bath. The reagents are then added at a constant volume ratio of 2:1 (diene:SCl_2) over a period of about 1 hour, and the flask is maintained in the cold bath for an additional 2 hours. The reaction mixture is then brought to room temperature by gradual warming overnight, heated briefly to 50° to dissolve all solids, cooled in an ice bath, and filtered. The residue is washed with hexane and air dried to give about 75 g (50%) of product, mp 100–101°. An additional 20 g may be obtained by concentration of the filtrate, filtration of the resulting solids, and recrystallization from methylene chloride.

2. *1,5-Cyclooctane Sulfide:* To 12.5 g (0.06 mole) of the dichlorosulfide in 150 ml of ether is added 1.2 g (0.03 mole) of lithium aluminum hydride in 60 ml of ether at a rate so as to maintain a gentle reflux (about 20 minutes). The mixture is allowed to stand overnight and is then cautiously treated with water to decompose the excess hydride. The mixture is mixed with fuller's earth (Floridin) and is filtered, and the filtrate is dried over anhydrous magnesium sulfate. Filtration of the solution and evaporation of the solvent (rotary evaporator) gives about 7 g of the colorless crystalline solid, mp 170–171°. It may be recrystallized from aqueous methanol, mp 172–173°.

VI. Methylation by Diazomethane

Diazomethane is a yellow gas that is toxic and explosive but that may be handled safely in solution in ether. It reacts immediately with an acid to liberate nitrogen and form the methyl ester. Its reaction with alcohols to form methyl esters requires catalysis by a Lewis acid. The procedures illustrate the use of this reagent as a methylating agent (see Chapter 17, Section III, for preparation of diazomethane).

A. METHYL ESTERS (9)

$$RCOOH + CH_2N_2 \rightarrow RCOOCH_3 + N_2\uparrow$$

The following operations are carried out in a hood behind a safety shield!

The acid (0.1 to 0.5 g) is dissolved or suspended in ether, and the ethereal diazomethane solution is added in small portions with swirling until the yellow color of diazomethane persists and nitrogen gas is no longer evolved. The solution is then warmed on a steam bath briefly to expel the excess reagent and the ether is evaporated to give the desired methyl ester. Examples are given in Table 7.2.

TABLE 7.2

	mp (°C)
1. Itaconic acid	162–164
Dimethyl itaconate	37–40
2. Diphenylacetic acid	148
Methyl diphenylacetate	60
3. Benzilic acid	150–153
Methyl benzilate	75
4. *p*-Nitrocinnamic acid	289
Methyl *p*-nitrocinnamate	161

B. Methyl Ethers: Catalysis by Aluminum Chloride (*10*)

$$R—OH + CH_2N_2 \xrightarrow[\text{Et}_2\text{O}]{\text{AlCl}_3} R—O—CH_3 + N_2\uparrow$$

An alcohol-free solution of diazomethane in ether is prepared as in Chapter 17, Section III. This solution is approximately 0.5 *M* in diazomethane and may be standardized by titration as follows: benzoic acid (0.6 g, approx. 0.005 mole) is weighed accurately into an Erlenmeyer flask and suspended in 5 ml of ether. The diazomethane solution (approx. 5 ml) is added from a buret with swirling, care being taken that an excess of unreacted benzoic acid remains (the yellow color of the diazomethane should be completely discharged). The excess benzoic acid is now titrated with standard 0.2 *N* sodium hydroxide solution, and the concentration of diazomethane is calculated.

General Procedure

The alcohol (5 mmole) is dissolved in 5 ml of ether in a conical flask and approx. 5 mg (0.38 mmole) of anhydrous aluminum chloride is added. The mixture is stirred with a magnetic stirring bar and cooled in an ice–salt bath to 0°. To the mixture is added dropwise the ethereal solution of diazomethane at a rate of about 2 ml per minute. The yellow color of the diazomethane disappears rapidly, and nitrogen is vigorously evolved. After the addition of about 4 mmole of diazomethane, the reaction may become sluggish. The addition of an additional 2 mg portion of anhydrous aluminum chloride will increase the reaction rate. The diazomethane solution is added until the yellow color persists for several minutes and is then allowed to stand at the cold temperature for 1 hour. (If the yellow color remains at the end of 1 hour, addition of 2 mg of aluminum chloride followed by dropwise addition of *absolute* ethanol will rapidly dispel the color.) The solution is now carefully washed with 50 ml of saturated sodium bicarbonate solution, followed by three water washes. After drying of the solution (anhydrous sodium sulfate) and removal of the solvent (rotary evaporator), the residue is purified by distillation or recrystallization as appropriate. Examples are given in Table 7.3.

TABLE 7.3

Alcohol	bp (mp) of methyl ether (°C)	Yield (%)
n-Octanol	173–174	83
Cyclohexanol	134	84
Cholesterol	(84 from acetone)	73
t-Amyl alcohol	86.3	49

C. METHYL ETHERS: CATALYSIS BY BORON TRIFLUORIDE (*10*)

The same general procedure as in the preceding case is employed. In place of the anhydrous aluminum chloride, boron trifluoride etherate (2 drops, approx. 0.8 mmole) is employed as the catalyst. The reaction technique and work-up are the same.

D. METHYL ETHERS: CATALYSIS BY FLUOROBORIC ACID

A detailed procedure for the preparation of cholestanyl methyl ether from cholestanol has been published (*11*) and a survey of the usefulness of the reagent has also appeared (*12*). There appears to be no particular advantage to this procedure over the more convenient alternatives given above.

VII. Oxymercuration: A Convenient Route to Markovnikov Hydration of Olefins

The acid catalyzed hydration of olefins is frequently attended by decomposition or rearrangement of the acid-sensitive substrate. A simple and mild procedure for the Markovnikov hydration of double bonds has recently been devised by Brown and co-workers (*13*). This reaction, which appears to be remarkably free of rearrangements, initially involves the addition of mercuric acetate to the double bond to give the 1,2-

$$CH_3(CH_2)_3CH{=}CH_2 \xrightarrow{\text{Hg(OAc)}_2} CH_3(CH_2)_3\underset{\underset{OH}{|}}{CH}{-}CH_2{-}HgOAc$$

hydroxymercury acetate. Treatment of the solution by sodium borohydride cleaves the organomercury bond giving the alcohol. Another interesting feature of the reaction is its stereoselectivity, particularly in the case of norbornene and related compounds (*14, 15*). Application of the reaction to norbornene gives 99.8% of the exo alcohol in about 85% yield. This and other examples are given in the procedures.

General Procedure (13)

$$R{-}CH{=}CH_2 \xrightarrow{\text{Hg(OAc)}_2} R\underset{\underset{OH}{|}}{CH}{-}CH_2{-}HgOAc \xrightarrow{\text{NaBH}_4} R\underset{\underset{OH}{|}}{CH}{-}CH_3$$

A 100-ml round-bottom flask equipped with a magnetic stirrer is charged with 3.19 g (0.01 mole) of mercuric acetate followed by 10 ml of water. The mercuric acetate dissolves to give a clear solution. Tetrahydrofuran (10 ml) is added to the solution forming a yellow suspension. The olefin (0.010 equivalents) is added to the mixture, which is stirred at room temperature until the reaction mixture becomes colorless and clear (usually in 1–10 minutes; see examples). Stirring is continued for an additional 10 minutes. Sodium hydroxide (10 ml of 3 M solution) is added followed by 10 ml of a solution, which is 0.5 M in sodium borohydride and 3 M in sodium hydroxide. The reduction is almost instantaneous. The mercury is allowed to settle, and the water layer is saturated with sodium chloride. The tetrahydrofuran (upper) layer is separated, dried with anhydrous sodium sulfate then filtered, and the solvent is removed (rotary evaporator). Distillation of the residue affords a high yield of the desired alcohol. The reaction scale may be increased to any level, but care must be taken to control the exothermic reactions (both steps) by the use of a cooling bath. Examples are given in Table 7.4. (*13*).

TABLE 7.4

Olefin	Reaction time for addition (minutes)	Product	bp of product (1 atm) (°C)
1-Hexene	10	2-Hexanol	139–140
3,3-Dimethyl-1-butene	20	3,3-Dimethyl-2-butanol	120–121
Cyclopentene	60	Cyclopentanol	140
Cyclohexene	11	Cyclohexanol	160
Styrene	5	1-Phenylethanol	203

STEREOSELECTIVE OXYMERCURATION-DEMERCURATION OF NORBORNENE (15)

The procedure given above is applied to norbornene. However, the formation of the alcohol is accompanied by formation of moderate amounts of the acetate. Therefore, the dried tetrahydrofuran solution of the alcohol–acetate mixture is treated with 0.4 g (0.01 mole) of lithium aluminum hydride dissolved in 10 ml of THF. The excess hydride is decomposed by careful addition of water, followed by filtration, drying of the organic solution, and evaporation of the solvent. The residue is almost pure (>99.8%) *exo*-2-norborneol. It may be purified by direct distillation, bp 178–179°/1 atm, crystallizing slowly on cooling, mp 127–128°.

VIII. Esterification of Tertiary Alcohols

Normal Fischer esterification of tertiary alcohols is unsatisfactory because the acid catalyst required causes dehydration or rearrangement of the tertiary substrate. Moreover, reactions with acid chlorides or anhydrides are also of limited value for similar reasons. However, treatment of acetic anhydride with calcium carbide (or calcium hydride) followed by addition of the dry tertiary alcohol gives the desired acetate in good yield.

A. *t*-BUTYL ACETATE (16)

$$(CH_3)_3COH + (CH_3CO)_2O \xrightarrow{CaC_2} (CH_3)_3COOCCH_3 + CH_3COOH$$

A 500-ml, three-necked, round-bottom flask is fitted with a condenser, a thermometer, and a mechanical stirrer; all openings are protected by drying tubes. In the flask is placed a mixture of 61.2 g (0.61 mole) of acetic anhydride and 27 g (about 0.4 mole) of finely pulverized calcium carbide, and the mixture is refluxed for 2 hours. After cooling to 70°, 29.6 g (0.4 mole) of dry *t*-butyl alcohol is added and the mixture is refluxed for 4 days with continuous stirring. The mixture is cooled and the thick slurry is decomposed by careful addition to ice, followed by steam distillation. The ester layer is separated, washed with sodium carbonate solution and water, and dried over anhydrous magnesium sulfate. Filtration, followed by fractional distillation, gives the desired ester almost without a forerun, bp 95–96.5°, yield 80–90%.

B. Pinacol Diacetate (*16*)

$$(CH_3)_2\underset{\underset{HO}{|}}{C}-\underset{\underset{OH}{|}}{C}(CH_3)_2 + (CH_3CO)_2O \xrightarrow{CaC_2} (CH_3)_2\underset{\underset{AcO}{|}}{C}-\underset{\underset{OAc}{|}}{C}(CH_3)_2 + CH_3COOH$$

Acetic anhydride (75 ml) and pulverized calcium carbide (15 g) are refluxed for 60 minutes in an apparatus as described above. Sufficient benzene is then added to reduce the boiling temperature to 110–112°. After cooling, 14.4 g of pinacol is introduced, and the mixture is refluxed with stirring for 30 hours. The cooled mixture is then stirred into ice, extracted into ether and washed with sodium carbonate solution. The solution is dried, then filtered, and the ether is evaporated. The residue is clarified with Norit and recrystallized from benzene–petroleum ether or methanol, mp 67–68°.

IX. Ketalization

Conversion of a ketone into a ketal is frequently employed where protection of the group is required (*17*). An example of the use of the process is the preparation of 4-ketocyclohexanol. The example presented in the procedure is typical. The product of

this procedure may be converted via hydrogenolysis into cyclohexyloxyethanol (Chapter 2, Section II).

CYCLOHEXANONE KETAL (1,4-DIOXASPIRO[4.5]DECANE) (*18*)

A mixture of cyclohexanone (11.8 g, 0.12 mole), ethylene glycol (8.2 g, 0.13 mole), *p*-toluenesulfonic acid monohydrate (0.05 g), and 50 ml of benzene is placed in a 250-ml round-bottom flask fitted with a water separator and a condenser (drying tube). The flask is refluxed (mantle) until the theoretical amount of water (approx. 2.2 ml) has collected in the separator trap. The cooled reaction mixture is washed with 20 ml of 10% sodium hydroxide solution followed by five 10-ml washes with water, dried over anhydrous potassium carbonate, and filtered. The benzene is removed (rotary evaporator) and the residue is distilled, affording 1,4-dioxaspiro[4.5]decane, bp 65–67°/13 mm, n_D^{25} 1.4565–1.4575, in about 80% yield.

X. Half-Esterification of a Diol

The general procedures for the preparations of half-esters of diols are not usually very efficient in that both unreacted diol and diester are present along with the desired half-ester. Unfortunately, the present example is no exception but is included as an illustration of the type of results that can be expected under these circumstances. The diol employed for this reaction is prepared in Chapter 5, Section II.

4-BENZOYLOXYCYCLOHEXANOL (*19*)

In a 250-ml three-necked flask fitted with a magnetic stirrer, a pressure-equalizing dropping funnel, and a thermometer is placed a solution of 1,4-cyclohexanediol (11.4 g, 0.10 mole), 35 ml of chloroform, and 27 ml of dry pyridine. The solution is cooled in an ice bath to 0–5° and is maintained below 5° throughout the addition. A solution of benzoyl chloride (14 g, 0.10 mole) in 30 ml of dry chloroform is added with stirring at a rate so as to keep the temperature below 5° (approx. 40 minutes). After completion of the addition, the mixture is allowed to come to room temperature and stand overnight. The chloroform solution is washed four times with 50-ml portions of water, once with 50 ml of 5% sulfuric acid solution, and finally with saturated sodium chloride solution. The chloroform solution is then dried (sodium sulfate), and the solvent is removed. Fractionation of the residue gives a cis and trans mixture of 4-benzoyloxycyclohexanol, bp 175–178°/0.2 mm, as a very viscous oil, yield about 55%.

XI. Substitution on Ferrocene

Since the discovery in 1951 of ferrocene by Kealy and Pauson (*20*), compounds containing organic groups bonded with π-electrons to transition metals have been prepared frequently and studied extensively. They exhibit unusual stability for organo-metallic compounds, and present several interesting challenges to the synthetic chemist. Such compounds are spoken of as possessing "aromatic stability," and yet straight-forward substitution reactions by electrophilic reagents are irregularly successful. Ferrocene, for example, despite its ability to undergo Freidel-Crafts type acylations, cannot be successfully nitrated by any of the nitronium ion producing systems. Nitro-ferrocene must be produced by the indirect route of metallation followed by treatment with an alkyl nitrate.

NITROFERROCENE VIA FERROCENYL LITHIUM (*21*)

The reaction is carried out under argon in a 2-liter three-necked flask fitted with a mechanical stirrer, reflux condenser, 250-ml pressure-equalizing addition funnel, and gas inlet and outlet. After purging with argon, the flask is charged with a solution of 89 g (0.48 mole) of ferrocene in 1 liter of dry tetrahydrofuran. The solution is next heated to 45°, and there is added dropwise with stirring, 155 ml (0.29 mole) of an *n*-butyllithium solution (15% in heptane–pentane, 2:1, Foote Mineral Co.) during a period of 75 minutes. The resultant solution is maintained at 45° for an additional 2 hours, then is cooled to −77° by means of an external Dry Ice–chloroform bath.

To the cooled solution is added dropwise with stirring, a solution of 64 g (0.48 mole) of *n*-amyl nitrate in 100 ml of dry tetrahydrofuran during a period of 1 hour. The solution is finally allowed to warm to room temperature overnight. Work-up is carried out by adding an excess of ice–acetic acid, diluting with water, and extracting the aqueous solution several times with ether. The ether extracts are washed thoroughly with water, dried over anhydrous sodium sulfate, filtered, and evaporated to dryness to yield 85 g (83% recovery) of a mixture of unchanged ferrocene, nitroferrocene, and 1,1′-dinitroferrocene.

Chromatography of the mixture on Activity I basic alumina yields, successively, ferrocene; nitroferrocene (11.6 g, 17%), mp 129.5–130.5° (recrystallized from pentane); and 1,1′-dinitroferrocene (0.130 g), mp 207–209° (dec).

XII. Demethylation of Aryl Methyl Ethers by Boron Tribromide

Boron tribromide reacts with ethers by the sequence shown (22). The boron complex

$$R—O—R + BBr_3 \rightarrow (RO)_3B + 3\ R—Br$$

$$(RO_3)B + 3\ H_2O \rightarrow 3\ ROH + H_3BO_3$$

can readily be hydrolized with water, yielding a mixture of alcohols and alkyl bromides. When one of the groups is an aryl group, the reaction proceeds in only one sense to give phenols and alkyl bromides in good yield (23). Since the reaction may be carried out at low temperatures, the procedure is usually the method of choice for demethylation of sensitive aryl methyl ethers.

General Procedure and Examples (24)

$$Ar—OCH_3 + BBr_3 \rightarrow (Ar—O)_3B + CH_3Br$$
$$\downarrow {\scriptstyle H_2O}$$
$$\longrightarrow Ar—OH$$

The amount of boron tribromide employed should equal 1 equivalent for each methoxyl group present, plus 1 equivalent for each other basic group present (—CN, —COOH, etc.). The boron tribromide is dissolved in 5 to 10 times its volume of methylene chloride and is placed in a suitable round-bottom flask fitted with a pressure-equalizing dropping funnel (drying tube) and a magnetic stirrer. The dropping funnel is charged with a solution of the aryl methyl ether (1 equivalent) in 30 to 40 times its weight of methylene chloride, and this solution is added to the stirred flask at room temperature over about 1 hour. The mixture is then allowed to stand overnight. (If the ether is sensitive, the addition may be carried out with the stirred flask cooled to −80° in a Dry Ice cooling bath followed by gradual warming to room temperature overnight.) Water is then added to hydrolyze the boron complexes and the mixture is extracted with ether. The ether extract is extracted with 2 N sodium hydroxide and the alkaline extract acidified (dilute hydrochloric acid). Extraction with ether followed by drying (sodium sulfate) and removal of the ether affords the crude product. Examples are given in Table 7.5 (24).

TABLE 7.5

Compound	Starting temp. (°C)	Moles BBr_3:compound	mp of product (°C)	Yield (%)
1-Methoxynaphthalene	−80	1:1	94–96	13.5
2-Methoxynaphthalene	−80	1:1	122–124	67
4-Methoxybenzoic acid	−80	2.3:1	214–215	92
3,4-Dimethoxybenzoic acid	−80	2.1:1	200–202	90
3,5-Dimethoxybenzoic acid	20	2.2:1	236–238	31
3,3′-Dimethoxybiphenyl	−80	1.7:1	124	77

REFERENCES

1. A. C. Cope and E. Ciganek, *Org. Syn.* **39**, 40 (1959).
2. D. Todd, *Org. React.* **4**, 378 (1948).
3. Huang-Minlon, *J. Amer. Chem. Soc.* **68**, 2487 (1946).
4. W. G. Dauben, J. B. Rogan, and E. J. Blanz, *J. Amer. Chem. Soc.* **76**, 6384 (1954).
5. A. I. Meyers, W. N. Beverung, and G. Garcia-Munoz, *J. Org. Chem.* **29**, 3427 (1964).
6. W. G. Dauben, E. C. Martin, and G. J. Fonken, *J. Org. Chem.* **23**, 1205 (1958); A. S. Hussey J. Sauvage, and R. H. Baker, *J. Org. Chem.* **26**, 256 (1961).
7. C. R. Hauser and D. S. Hoffenberg, *J. Org. Chem.* **20**, 1448 (1955).
8. E. D. Weil, K. J. Smith, and R. J. Gruber, *J. Org. Chem.* **31**, 1669 (1966).
9. See, for example, A. I. Vogel, "Practical Organic Chemistry," 3rd ed., p. 973. Longmans, London, 1956.
10. E. Muller, R. Heischkeil, and M. Bauer, *Ann. Chem.* **677**, 55 (1964).
11. M. Neeman and W. S. Johnson, *Org. Syn.* **41**, 9 (1961).
12. M. Neeman, M. C. Caserio, J. D. Roberts, and W. S. Johnson, *Tetrahedron* **6**, 36 (1959).
13. H. C. Brown and P. Geoghegan, *J. Amer. Chem. Soc.* **89**, 1522 (1967).
14. H. C. Brown and W. J. Hammar, *J. Amer. Chem. Soc.* **89**, 1524 (1967).
15. H. C. Brown, J. H. Kawakami, and S. Ikegami, *J. Amer. Chem. Soc.* **89**, 1525 (1967).
16. R. V. Oppenauer, *Monatsh. Chem.* **97**, 62 (1966).
17. For example, W. S. Johnson *et al.*, *J. Amer. Chem. Soc.* **78**, 6300 (1956).
18. R. A. Daignault and E. L. Eliel, *Org. Syn.* **47**, 37 (1967).
19. E. R. H. Jones and F. Sondheimer, *J. Chem. Soc.*, p. 615 (1949).
20. T. J. Kealy and P. L. Pausen, *Nature* **68**, 1039 (1951).
21. R. E. Bozak, personal communication.
22. F. L. Benton and T. E. Dillon, *J. Amer. Chem. Soc.* **64**, 1128 (1942).
23. J. F. W. McOmie and M. L. Watts, *Chem. Ind.* (*London*), p. 1658 (1963).
24. J. F. W. McOmie, M. L. Watts, and D. E. West, *Tetrahedron* **24**, 2289 (1968).

II

SKELETAL MODIFICATIONS

8

The Diels-Alder Reaction

One of the most powerful tools for the formation of cyclic molecules is the Diels-Alder reaction (*1*). The reaction generally involves the combination of a diene with a "dienophile" according to the diagram. There are surprisingly few limitations on the

character of either fragment, although the presence of electron withdrawing substituents at A and B enhances the reaction rate. The number of examples of this reaction that have been studied is vast, and the procedures given here are typical.

I. 3,6-Diphenyl-4,5-cyclohexenedicarboxylic Anhydride

Maleic anhydride is a convenient dienophile because of its rapid reaction with most dienes as well as its stability and ease in handling (although it is poisonous). The diene for this reaction, 1,4-diphenyl-1,3-butadiene, is readily prepared by the Wittig reaction with benzyltriphenylphosphonium chloride and cinnamaldehyde (Chapter 11, Section I).

3,6-DIPHENYL-4,5-CYCLOHEXENEDICARBOXYLIC ANHYDRIDE (*2*)

A 100-ml round-bottom flask is charged with a mixture of xylene (25 ml), 1,4-diphenyl-1,3-butadiene (2.3 g), and finely powdered maleic anhydride (1.1 g). (Larger quantities may be used if desired as long as the reagents are equimolar.) The flask is fitted with a condenser, the solution is refluxed for 15 minutes, the reaction mixture is allowed to cool to room temperature, and the adduct is collected. Recrystallization from xylene or another suitable solvent affords the pure product, mp 203–205°.

II. Reactions with Butadiene

1,3-Butadiene is an important reagent in Diels-Alder reactions, but it may be troublesome to work with at elevated temperatures because of its high vapor pressure. Two procedures are given in which gaseous butadiene is employed as the diene in reactions at moderate pressure. A third procedure describes the use of 3-sulfolene as a convenient source of 1,3-butadiene generated *in situ*. Upon heating (110–130°), 3-sulfolene decomposes to give 1,3-butadiene and sulfur dioxide. In the presence of a

suitable dienophile, the Diels-Alder reaction occurs.

A. 4,5-CYCLOHEXENEDICARBOXYLIC ANHYDRIDE (3)

Benzene (80 ml) is placed in a suitable pressure vessel (soft drink bottle or hydrogenation bottle) and chilled to 5°. The bottle is weighed, and a gas dispersion tube connected to a cylinder of butadiene is immersed in the benzene. Butadiene is introduced into the flask with continued cooling until a total of 32 g has been transferred. Pulverized maleic anhydride (50 g) is added to the bottle, which is then capped or stoppered with a stopper wired in place. The bottle is allowed to stand at room temperature for 12 hours, then is heated (behind a safety shield) to 100° for 5 hours. The bottle is cooled, then opened, and the contents are transferred to an Erlenmeyer flask. The mixture is heated to boiling, and petroleum ether is added until there is a slight turbidity. After cooling, the product is collected, mp 101–103° (yield 90%).

B. *cis*-4,5-CYCLOHEXENEDICARBOXYLIC ACID

4,5-cyclohexenedicarboxylic anhydride is refluxed in water for 2 hours. The solution is cooled and the crystals of product are collected. Recrystallization from water may be carried out, if desired. The product has mp 164–166°.

C. 4,5-CYCLOHEXENEDICARBOXYLIC ANHYDRIDE: 3-SULFOLENE AS A SOURCE OF BUTADIENE (4)

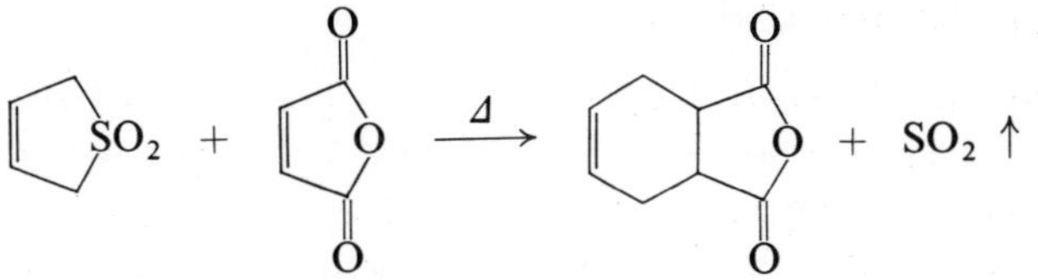

Note: The generation of sulfur dioxide requires that the following reaction be carried out in a hood.

Twenty-five grams (0.212 mole) of 3-sulfolene, and 15.0 g (0.153 mole) of pulverized maleic anhydride are added to a dry 250-ml flask fitted with a condenser. Boiling chips and 10 ml of dry xylene are added. The mixture is swirled gently for a few minutes to effect partial solution, then gently heated until an even boil is established. During the first 5–10 minutes, the reaction is appreciably exothermic, and care must be exercised to avoid overheating.

After 25–30 minutes the flask is allowed to cool. (Cooling should be no longer than about 5 minutes otherwise the product will separate as a hard cake difficult to redissolve.) About 150 ml of benzene and 2 g of Norit are then added to the flask (the carbon serves not only to decolorize the product, but also as a necessary filter aid). The suspension is brought to boiling on a steam bath and is filtered hot into a 250-ml flask. A fresh boiling stone is added and the filtrate is simmered until the product redissolves. Petroleum ether is added with swirling until a slight turbidity persists. The solution is allowed to cool, and the product is collected, mp 104–106° (yield 82–90%). (The *cis*-dicarboxylic acid may be prepared by refluxing the anhydride with water as above.)

THE REACTION OF QUINONE WITH BUTADIENE (5)

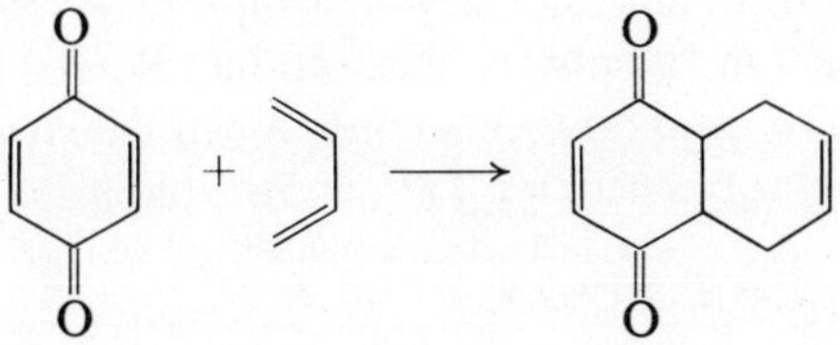

p-Benzoquinone (practical grade, 5 g) is placed in a 100-ml round-bottom flask, 35 ml of benzene is added, and the mixture is cooled to 5° in an ice bath. By means of a gas dispersion tube, 3.1 g (5 ml) of butadiene is introduced and the flask is tightly closed with a stopper wired in place. The reaction mixture is brought to room temperature and is allowed to stand for 2–3 weeks with occasional swirling during the first week to facilitate solution of the quinone. After the reaction period is over, the dark mixture is treated with Norit, filtered through celite, and the solvent is removed under reduced pressure. The residue is recrystallized from petroleum ether with a little benzene, mp 52–54° (yield 80–93 %).

III. Catalysis by Aluminum Chloride

The Diels-Alder reaction was thought for many years to be only slightly influenced by catalysts. However, in 1960, Yates and Eaton (6) clearly demonstrated that with certain dienophiles, the presence of a molar equivalent of aluminum chloride can cause a remarkable acceleration of the reaction. Providing the diene is not polymerized (7) or otherwise destroyed by the catalyst, the modification can be fruitfully employed to carry out the reaction at lower temperature and for shorter times.

General Procedure (6)

In a 250-ml flask equipped with a magnetic stirrer is placed 160 ml of methylene chloride. The diene (0.01 mole) and the dienophile (0.01 mole) are added, and stirring is begun. In one batch, 0.01 mole of anhydrous aluminum chloride is added to the flask, which is then closed with a drying tube, and stirring is continued until the reaction is complete (2–120 minutes). The reaction mixture is then poured onto ice (200 g), mixed thoroughly, and the methylene chloride layer is separated. The methylene chloride solution is now washed with two 100-ml portions of water, dried (sodium sulfate), and the solvent is removed. The residue is recrystallized to give the desired adduct in a good state of purity.

Examples

1. Anthracene and maleic anhydride with aluminum chloride give the adduct quantitatively in $1\frac{1}{2}$ minutes. The product is recrystallized from ethyl acetate, mp 262–263°.

2. Anthracene, dimethyl fumarate, and aluminum chloride give the adduct in 2 hours. When 2 molar equivalents of aluminum chloride are used, the reaction is complete in 5 minutes. The product is recrystallized from methanol, mp 108–109°.

3. Equimolar amounts of anthracene, p-benzoquinone, and aluminum chloride give the faintly yellow adduct in 15 minutes. The product is unstable to heat turning yellow at 207°, turning red at 210°, and slowly charring. When 2 molar equivalents of anthracene are used, the bis adduct is obtained, mp 230°, unobtainable in the absence of the catalyst.

CATALYSIS OF THE BUTADIENE REACTION

Note: The following reaction should be carried out in a hood.

1. *Reaction with Methyl Acrylate (8):* In a 500-ml three-necked flask equipped with a pressure-equalizing addition funnel, a magnetic stirrer, a condenser, and a gas dispersion tube immersed in the solvent, is placed 250 ml of benzene followed by 6.8 g (0.05 mole) of anhydrous aluminum chloride. The flask is warmed to about 50° and a solution of 43.2 g (0.5 mole) of methyl acrylate in 50 ml of benzene is added over 10 minutes with stirring. The aluminum chloride rapidly dissolves to form a clear yellow solution. Butadiene (excess, a total of about 1.6 mole) is bubbled through the stirred solution over a period of about 4 hours, the temperature of the flask being maintained at 50–60° by gentle heating. The cooled reaction mixture is then poured into ice water, and the benzene layer is separated and washed twice with water. After drying (sodium sulfate) and removal of the solvent (rotary evaporator), the residue is distilled to give about 60 g (85%) of methyl 3-cyclohexene-1-carboxylate, bp 73–74°/20 mm, n_D^{20} 1.4602.

Note: The following reaction should be carried out in a hood.

2. *Reaction with Acrylonitrile (9):* As in the preceding case, a mixture of 61 g (0.45 mole) of aluminum chloride in 300 ml of benzene is heated to 60°, and a solution of 26.6 g (0.5 mole) of acrylonitrile in 100 ml of benzene is added. Butadiene (0.9 mole) is bubbled into the stirred and heated solution over a period of 4 hours, and the reaction mixture is worked up as above. Distillation gives 3-cyclohexene-1-carbonitrile, bp 80–87°/20 mm, n_D^{20} 1.4742, in about 85% yield.

IV. Generation of Dienes from Diones

Cyclic 1,3-diacetoxy-1,3-dienes can be generated *in situ* from cyclic 1,3-diketones under the influence of isopropenyl acetate. The dienes then undergo Diels-Alder reactions with maleic anhydride giving as products 1-acetoxybicycloalkane dicarboxylic anhydride derivatives (*10*). The procedure is also successful with cyclic 1,2- and 1,4-diketones as well as cyclic α,β-unsaturated ketones. The products, after hydrolysis to

the corresponding diacids, may be subjected to lead tetraacetate catalyzed oxidative decarboxylation (bisdecarboxylation) to give bicycloalkene derivatives (Chapter 1, Section X).

A. *endo*-1,3-DIACETOXY-8,8-DIMETHYLBICYCLO[2.2.2]OCT-2-ENE-5,6-DICARBOXYLIC
ANHYDRIDE (*10*)

In a 500-ml round-bottom flask is placed a mixture of 25 g (0.178 mole) of dimedone, 21.8 g (0.22 mole, 25% excess) of pulverized maleic anhydride, 0.1 g of *p*-toluenesulfuric acid, and 150 ml of isopropenyl acetate. The mixture is refluxed for 72 hours, then cooled, and the acetone is removed at room temperature on a rotary evaporator. The resulting solution is cooled to −20° in a Dry Ice bath, whereupon the product crystallizes. It is collected by filtration to yield the crude product in about 80% yield. Recrystallization from hexane–ethyl acetate and decolorization by Norit gives colorless crystals, mp 164–166°.

B. *endo*-1-ACETOXY-8,8-DIMETHYLBICYCLO[2.2.2]OCT-3-ONE-5,6-DICARBOXYLIC ACID
(*10*)

The anhydride prepared above is refluxed in water for 4–5 hours. The solution is cooled and extracted six times with small portions of ether (or better, continuously extracted with ether). Drying (sodium sulfate) of the ethereal solution, followed by evaporation of the solvent gives the desired acid, mp 146–149° (dec).

C. *endo*-1,3-DIACETOXYBICYCLO[2.2.2]OCT-2-ENE-5,6-DICARBOXYLIC ANHYDRIDE (*10*)

The reaction of 20 g (0.177 mole) of 1,3-cyclohexanedione (Chapter 5, Section II) with 21.8 g (0.22 mole) of maleic anhydride and 0.1 g of *p*-toluenesulfonic acid in 150 ml of isopropenyl acetate is conducted as described above to give about 70% of the recrystallized product, mp 156–159°.

D. *endo*-1-ACETOXYBICYCLO[2.2.2]OCT-3-ONE-5,6-DICARBOXYLIC ACID (*10*)

The corresponding anhydride is refluxed in water for 5 hours, and the product is isolated as in the preceding case, mp 148–151° (dec).

E. *endo*-2-ACETOXYBICYCLO[2.2.2]OCT-2-ENE-5,6-DICARBOXYLIC ANHYDRIDE (*10*)

The reaction of 17 g (0.177 mole) of 2-cyclohexene-1-one under the conditions previously described gives about 29 g of tan solid on cooling. Several recrystallizations may be necessary to remove a small amount of isomer formed. The pure product has mp 138–141°.

V. Reactions with Cyclopentadiene

The use of cyclopentadiene as a diene provides a route to the [2.2.1]bicycloheptane skeleton, which is of considerable theoretical interest. Cyclopentadiene, however, exists as its Diels-Alder dimer at room temperature and must be "cracked" thermally to obtain the monomer. Below are given the procedures for cracking the dimer and for a convenient addition reaction of the monomer.

A. CYCLOPENTADIENE FROM THE DIMER (11)

A 100-ml flask containing 20 ml of dicyclopentadiene is set up for fractional distillation at atmospheric pressure, the receiver being chilled in an ice bath. The flask is gently heated and the contents are allowed to come to a brisk reflux, whereupon the monomer begins to distil at 40–42°. Distillation is continued at a rate so as to keep the distillate temperature below 45°. When the pot residue is about 5 ml in volume, the heating is discontinued, and about a gram of anhydrous calcium chloride is added to the distillate to ensure its dryness. Although cyclopentadiene should be prepared fresh for each use, it may be stored for short periods at 0° without appreciable dimerization occurring.

B. *endo*-BICYCLO[2.2.1]HEPT-2-ENE-5,6-DICARBOXYLIC ANHYDRIDE (11)

Pulverized maleic anhydride (6 g, 0.061 mole) is dissolved in 20 ml of ethyl acetate by gentle heating on a steam bath. Petroleum ether (20 ml) is added slowly to the solution, which is then cooled in an ice bath. To the cold solution is added 4.8 g (6 ml, 0.073 mole) of cyclopentadiene, and the resulting solution is swirled until the exothermic reaction subsides and the product separates. Recrystallization may be carried out in the reaction solvent by heating until dissolution occurs (steam bath) followed by slow cooling. The product has mp 164–165°, yield about 80%.

The dicarboxylic acid is prepared as in Chapter 8, Section I by heating the anhydride in water until dissolution is complete. *endo*-Bicyclo[2.2.1]hept-2-ene-5,6-dicarboxylic acid has mp 180–182°.

REFERENCES

1. M. C. Kloetzel, *Org. React.* **4**, 1 (1948); H. L. Holms, *Org. React.* **4**, 60 (1948); L. W. Butz and A. W. Rytina, *Org. React.* **5**, 136 (1949).
2. W. Kemp, "Practical Organic Chemistry," p. 93. McGraw-Hill, London, 1967.
3. L. F. Fieser and F. C. Novello, *J. Amer. Chem. Soc.* **64**, 802 (1942).
4. T. E. Sample and L. F. Hatch, *J. Chem. Educ.* **45**, 55 (1968).
5. E. E. van Tamelen, M. Shamma, A. W. Burgstahler, J. Wolinsky, R. Tamm, and P. E. Aldrich, *J. Amer. Chem. Soc.* **91**, 7315 (1969).
6. P. Yates and P. Eaton, *J. Amer. Chem. Soc.* **82**, 4136 (1960).
7. G. I. Fray and R. Robinson, *J. Amer. Chem. Soc.* **83**, 249 (1961).
8. T. Inukai and M. Kasai, *J. Org. Chem.* **30**, 3567 (1965).
9. T. Inukai and T. Kojima, *J. Org. Chem.* **31**, 2032 (1966).
10. C. M. Cimarusti and J. Wolinsky, *J. Amer. Chem. Soc.* **90**, 113 (1968).
11. R. M. Roberts, J. C. Gilbert, L. B. Rodewald, and A. S. Wingrove, "Modern Experimental Organic Chemistry," p. 145. Holt, Rinehart and Winston, New York, 1969.

9

Enamines as Intermediates

Certain ketones react reversibly with secondary amines to give enamines. By removal of water, the equilibrium can be driven to the right, and the enamine can be isolated.

$$\text{C=O} + NHR_2 \underset{}{\overset{H^+}{\rightleftharpoons}} \text{C}\text{--}NR_2 + H_2O$$

The β-position of enamines is highly nucleophilic and may react with alkyl halides, acyl chlorides or anhydrides, or with Michael addition substrates to give carbon–carbon bonds as shown in the examples (1).

I. Preparation of the Morpholine Enamine of Cyclohexanone

The preparation of enamines of unsubstituted cyclic ketones is convenient and unambiguous in that only a single product results. Moreover, such enamines have found wide application as means of chain extension or ring formation.

1-MORPHOLINO-1-CYCLOHEXENE (2)

A 250-ml round-bottom flask is charged with a mixture of cyclohexanone (14.7 g, 0.15 mole), morpholine (15.7 g, 0.18 mole), and p-toluenesulfonic acid monohydrate (0.15 g) in 50 ml of toluene. The flask is fitted with a water separator and a condenser and is brought to reflux (mantle). The separation of water begins immediately and the theoretical amount (2.7 ml) is obtained in about 1 hour. Without further treatment, the reaction mixture may then be distilled. After removal of the toluene at atmospheric pressure, the product is obtained by distillation at reduced pressure, bp 118–120°/10 mm, n_D^{25} 1.5122–1.5129, in about 75% yield.

For many reactions, the enamine may be used without distillation. The cooled toluene solution is washed with bicarbonate followed by water. After drying, toluene is removed (rotary evaporator) and the crude enamine employed directly.

In general, enamines are sensitive to moisture, and this fact should be kept in mind if the enamine is not to be used immediately. If the enamine is to be stored for any length of time, refrigeration is recommended. Yellowing may occur on long storage, but this change appears to have little effect on the outcome of subsequent reactions.

II. Acylation of Enamines

In the acylation of enamines, the weakly basic acylated enamine does not absorb the acid formed. Consequently, one must employ 2 equivalents of the enamine or use a second tertiary amine to absorb the acid liberated. In the procedure, triethylamine is employed to absorb the hydrochloric acid.

2-ACETYLCYCLOHEXANONE (3)

In a 500-ml round-bottom flask is placed a solution of 25 g of the enamine and 20 ml of triethylamine in 175 ml of dry chloroform. Slowly, with swirling, a solution of 15 ml of acetyl chloride in 75 ml of dry chloroform is added, and the solution is refluxed for 2 hours (mantle).

To the cooled reaction mixture, water (40 ml) and concentrated hydrochloric acid (40 ml) are added, and the mixture is refluxed for 2 hours to hydrolyze the acetylated enamine.

The chloroform and aqueous acid layers are separated, and the chloroform layer is washed twice with 100-ml portions of water, these water washings being combined with the original aqueous acid layer. The combined aqueous layers are neutralized with 6 N sodium hydroxide solution and rendered *just* acid to litmus with hydrochloric acid. This solution is extracted twice with 50-ml portions of chloroform, which are added to the original chloroform layer. The combined chloroform layers are dried (calcium chloride) and the chloroform is evaporated. The residue is distilled under reduced pressure, bp 83–84°/6 mm, 120–130°/15 mm, yield 10–15 g.

III. Enamines as Michael Addition Reagents

The Michael reaction with enamines is exemplified in this procedure. In a second (spontaneous) step of the reaction, an aldol-type condensation occurs resulting in cyclization. Finally, the morpholine enamine of the product forms and is hydrolized by the addition of water to yield a mixture of octalones, which is separated by fractional crystallization. $\Delta^{1(9)}$-Octalone-2 can be reduced by lithium in anhydrous ammonia to the saturated *trans*-2-decalone (Chapter 3, Section III).

$\Delta^{1(9)}$-OCTALONE-2 (4)

1. $\Delta^{1(9)}$-CH$_2$=CHCOCH$_3$
2. H$_2$O

1. *$\Delta^{1(9)}$-Octalone-2 and $\Delta^{9,10}$-octalone-2:* A 500-ml, three-necked, round-bottom flask is fitted with a mechanical stirrer, a condenser, a dropping funnel, and a heating mantle. The flask is charged with a solution of 25.5 g (0.15 mole) of 1-morpholino-1-cyclohexene in 150 ml of dry dioxane (or benzene). Stirring is begun and methyl vinyl ketone (11.2 g, 0.16 mole) is added over about 30 minutes. The solution is then refluxed for 4 hours, water (200 ml) is added, and refluxing is continued for 10–12 hours. The cooled solution is diluted with 250 ml of water and the mixture is extracted four times with 100-ml portions of ether. The combined ether extracts are washed three times with 60-ml portions of 3 *N* hydrochloric acid, twice with 30-ml portions of bicarbonate solution, once with water, once with saturated sodium chloride solution, and finally are dried over anhydrous magnesium sulfate. After filtration and removal of the ether (rotary evaporator), the residue is distilled through a short column affording about 14 g (60%) of the octalone mixture, bp 75–78°/0.2 mm, 101–103°/2 mm. The mixture contains 10–20% of the $\Delta^{9,10}$-isomer.

2. *$\Delta^{1(9)}$-Octalone-2:* An 8.5-g (0.058 mole) portion of the above octalone mixture is dissolved in 50 ml of 60–90° petroleum ether in a 125-ml Erlenmeyer flask and cooled in a Dry Ice–acetone bath for 1 hour. $\Delta^{1(9)}$-Octalone-2 crystallizes and is collected by suction filtration through a jacketed sintered-glass funnel, which is cooled with Dry Ice–acetone. The residue is washed with cold (−78°) petroleum ether, transferred rapidly to a clean 125-ml Erlenmeyer flask, and the crystallization and filtration steps are repeated. The residue, after the second filtration, is transferred to a small round-bottom flask, brought to room temperature (the solid melts), and distilled. By this procedure, about 5 g (34%) of purified $\Delta^{1(9)}$-octalone-2, bp 143–145°/15 mm, is obtained. The purified material contains 1–3% of the $\Delta^{9,10}$-isomer.

Distillation of the mother liquors affords a fraction enriched in $\Delta^{9,10}$-octalone-2, bp 143–145°/15 mm.

IV. Reactions of Enamines with β-Propiolactone

β-Propiolactone is subject to attack by enolate ions to give propionic acid derivatives of ketones. It may likewise react with nucleophilic enamines to give carboxyethylation according to the reactions. The morpholide is easily hydrolyzed to the corresponding acid.

2-KETOCYCLOHEXANEPROPIONIC ACID (5)

A mixture of 100 g (0.6 mole) of 1-morpholino-1-cyclohexene, 28.8 g (0.4 mole) of β-propiolactone, and 100 ml of chlorobenzene is placed in a 500-ml round-bottom flask fitted with a condenser (drying tube). The mixture is refluxed for 4 hours. The solvent and excess enamine are removed by distillation at aspirator pressure. (The residue may be distilled to afford the pure morpholide, bp 187–188°/1 mm, n_D^{22} 1.5090.) Basic hydrolysis may be carried out on the undistilled morpholide. To the crude amide is added 400 ml of 10% sodium hydroxide solution. The mixture is cautiously brought to reflux, and refluxing is continued for 2 hours. The cooled reaction mixture is made acidic (pH 4) and is extracted three times with ether. The combined ether extracts are washed twice with 5% hydrochloric acid solution and twice with water. The ethereal solution is dried (sodium sulfate), then filtered, and the solvent is removed (rotary evaporator). The residue may be recrystallized from petroleum ether–benzene, mp 64°.

V. Reactions of Enamines with Acrolein

A novel ring closure was discovered by Stork (6) in which the pyrrolidine enamine of a cycloalkanone reacts with acrolein. The scheme illustrates the sequence in the case of 1-pyrrolidino-1-cyclohexene, and the cyclopentane compound was found to undergo the reaction analogously. The procedure details the preparation of the bicyclo adduct and its cleavage to 4-cyclooctenecarboxylic acid.

A. 1-PYRROLIDINO-1-CYCLOHEXENE

The procedure is the same as that described in Chapter 9, Section I for the preparation of the morpholine enamine of cyclohexanone. In place of morpholine, pyrrolidine (12.8 g, 0.18 mole) is used. The product is collected at 114–115°/15 mm, n_D^{20} 1.5200.

B. 2-*N*-PYRROLIDYLBICYCLO[3.3.1]NONAN-9-ONE AND METHIODIDE (*6, 7*)

1. In a 500-ml three-necked flask equipped with a mechanical stirrer, a condenser, and a dropping funnel (drying tubes) is placed a solution of 15.1 g (0.01 mole) of 1-pyrrolo-dino-1-cyclohexene in 100 ml of dry dioxane. The solution is cooled to 0–5° in an ice bath, and acrolein (5.9 g, 0.0105 mole) is added with stirring over 20 minutes. The solution is then allowed to stir at room temperature overnight. Dioxane and excess acrolein are removed (rotary evaporator) and the residue is distilled at reduced pressure to give the product, bp 125–127°/0.5 mm, yield 75%.

2. To a solution of 11.8 g of 2-*N*-pyrrolidylbicyclo[3.3.1]nonan-9-one in 25 ml of dry ether is added 25 g of methyl iodide in one portion. The solution is allowed to stand at room temperature for 2 hours, then filtered to remove the product. To the filtrate is added 5 g of methyl iodide and after 5 hours at room temperature, solid is again collected. A third crop is similarly obtained. The combined solids (approx. 17 g) are recrystallized from acetone–ethanol to give about 16 g of the methiodide, mp 220–222°.

C. 4-Cyclooctene-1-carboxylic Acid (7)

A mixture of 17 g of the methiodide and 32 ml of a 40 % aqueous potassium hydroxide solution is heated with stirring in a flask fitted with a condenser. The heating bath should be kept at 125–130°, and the heating should be continued for 5 hours. The cooled reaction mixture is then diluted with 30 ml of water and washed twice with 25-ml portions of ether. The aqueous layer is cautiously acidified in the cold with concentrated hydrochloric acid to a pH of about 2 and then extracted five times with 25-ml portions of ether. The combined extracts are washed twice with 10% sodium thiosulfate solution and are dried (magnesium sulfate). Removal of the solvent followed by distillation affords about 3 g of 4-cyclooctene-1-carboxylic acid, bp 125–126°/1.1 mm. The product may solidify and may be recrystallized by dissolution in a minimum amount of pentane followed by cooling in a Dry-Ice bath. After rapid filtration, the collected solid has mp 34–35°.

REFERENCES

1. G. Stork, A. Brizzolara, H. Landesman, J. Szmuszkovicz, and R. Terrell, *J. Amer. Chem. Soc.* **85**, 207 (1963); J. Szmuszkovicz, *in* "Advances in Organic Chemistry: Methods and Results" (R. A. Raphael, E. C. Taylor, and H. Wynberg, eds.), Vol. 4. Wiley/Interscience, New York, 1963.
2. S. Hunig, E. Lucke, and W. Brenninger, *Org. Syn.* **41**, 65 (1961).
3. W. Kemp, "Practical Organic Chemistry," p. 130. McGraw-Hill, London, 1967.
4. R. I. Augustine and J. A. Caputo, *Org. Syn.* **45**, 80 (1965).
5. G. Schroll, P. Klemmensen, and S. Lawesson, *Acta Chem. Scand.* **18**, 2201 (1964).
6. G. Stork and H. K. Landesman, *J. Amer. Chem. Soc.* **78**, 5129 (1956); see also C. S. Foote and R. B. Woodward, *Tetrahedron* **20**, 687 (1964).
7. A. C. Cope and G. L. Woo, *J. Amer. Chem. Soc.* **78**, 5130 (1956).

10

Enolate Ions as Intermediates

In the presence of strong bases, carbonyl compounds form enolate ions, which may be employed as nucleophilic reagents to attack alkyl halides or other suitably electron-deficient substrates giving carbon–carbon bonds. (The aldol and Claisen condensations

$$RCH_2\overset{\|}{\underset{O}{C}}-X \xrightarrow{\text{Base}} R\overset{\ominus}{C}H\overset{\|}{\underset{O}{C}}-X \longleftrightarrow RCH=\overset{}{\underset{\overset{|}{O^{\ominus}}}{C}}-X$$

$$\xrightarrow{R'-X} RCH-\overset{\|}{\underset{O}{C}}-X$$
$$\underset{R'}{|}$$

$$\xrightarrow{R'-CHO} \begin{matrix} R-CH-\overset{\|}{\underset{O}{C}}-X \\ | \\ R'-CH \\ | \\ O^{\ominus} \end{matrix}$$

are examples of this well-known process.) Since the generation of the enolate requires strong base, the scope of the reaction is limited to those substrates that are not base sensitive. Acyl chlorides, for example, cannot be employed as substrates for attack by enolates since they react rapidly with base. (Acyl chlorides may be used in the acylation of enamines, however; cf. Chapter 9). Therefore, acylation via enolates normally requires the use of esters as substrates. Several examples of this and other enolate-type condensations are given (1).

I. Ketones as Enolates: Carbethoxylation of Cyclic Ketones

Several methods for the carbethoxylation of cyclic ketones via their enolates have been described. The methods differ primarily in the nature of the acylating agent employed. Thus, diethyl oxalate, diethyl carbonate, and ethyl diethoxyphosphinyl formate all react with cycloalkanone enolates under suitable conditions to give acylated products. As is apparent from the procedures, the latter two methods are considerably more convenient.

A. 2-CARBETHOXYCYCLOHEXANONE BY CONDENSATION WITH DIETHYL OXALATE (2)

A 500-ml, three-necked, round-bottom flask is fitted with a mechanical stirrer, a condenser (drying tube), and a dropping funnel. The flask is charged with 150 ml of absolute ethanol, and sodium (11.5 g, 0.5 g-atom) is added cautiously (stirring unnecessary) to prepare a solution of sodium ethoxide. The flask is cooled to 10° with stirring, whereupon a previously cooled (ice bath) mixture of cyclohexanone (49 g, 0.5 mole) and diethyl oxalate (73 g, 0.5 mole) is added over about 15 minutes. (Rapid stirring is advisable to prevent solidification of the mixture.) After the addition, stirring in the cold is continued for 1 hour followed by stirring at room temperature for 6 hours. The stirred flask is again cooled to 5–10° and a mixture of 14 ml of concentrated sulfuric acid and 110 g of ice is cautiously added to decompose the reaction mixture, the temperature being maintained at 5–10°. The solution is then diluted with water (approx. 1 liter), and the oily ethyl 2-ketocyclohexylglyoxalate separated. The aqueous phase is extracted four times with 100-ml portions of benzene and the extracts are combined with the original organic phase. The benzene solution is washed twice with water and is transferred in portions to a 250-ml round-bottom flask set up for distillation on a steam bath. Distillation is continued until benzene no longer comes over. The distillation setup is now heated with an oil bath under vacuum, and the forerun is collected below 105°/10–12 mm. The product is collected at 105–165°/10–15 mm (final bath temperature of about 200°), yield about 65 g (65%). The distillate is transferred to a 125-ml flask set up for distillation, and approx. 1 mg of iron powder and 0.5 g of finely ground soft glass are added. The mixture is distilled at 40 mm, the decarbonylated product being collected over the range 125–140° (bath temperature 165–175°). The crude 2-carbethoxycyclohexanone, about 50 g (59%), may be purified by distillation, bp 93–94°/1 mm.

B. 2-Carbethoxycyclooctanone by Condensation with Diethyl Carbonate (3)

A 500-ml, three-necked, round-bottom flask is equipped with a magnetic stirrer, a pressure-equalizing dropping funnel, and a condenser (drying tube). Sodium hydride (50% dispersion in mineral oil, 10.2 g, 0.21 mole) is added and the mineral oil is removed by four washes with 25-ml portions of benzene, the benzene being removed by pipet after the hydride settles. Finally, a solution of 17.7 g (0.15 mole) of diethyl carbonate in 100 ml of benzene is placed in the flask and the mixture is brought to reflux. A solution of cyclooctanone (9.5 g, 0.075 mole) in 25 ml of benzene is added dropwise to the refluxing mixture over a period of about 1 hour followed by an additional reflux period of about 20 minutes (the hydrogen should have ceased to be evolved).

To the cooled (room temperature) reaction mixture, glacial acetic acid (15 ml) is added dropwise with stirring (formation of pasty solid), followed by 50 ml of ice-cold water (dissolution of the solid). The benzene layer is separated, the aqueous layer is extracted three times with 25-ml portions of benzene, and the combined benzene extracts are washed three times with 25-ml portions of cold water. Benzene is removed by distillation at atmospheric pressure, and excess diethyl carbonate is removed by distillation under aspirator pressure. The residue is distilled under vacuum, affording 2-carbethoxycyclooctanone, bp 85–87°/0.1 mm, n_D^{25} 1.4795–1.4800, about 14 g (94%).

C. 2-Carbethoxycyclohexanone by Condensation with Ethyl Diethoxyphosphinyl Formate

1. EtOOC—Cl + (EtO)$_3$P $\longrightarrow$ EtOOC—P(OEt)$_2$ + Et—Cl
                                              ‖
                                              O

2.

Note: Ethyl chloride is generated in the following reaction, which should therefore be carried out in a hood.

1. *Ethyl Diethoxyphosphinyl Formate (4):* A 250-ml three-necked flask is fitted with a dropping funnel, a thermometer, a condenser, and a magnetic stirrer. The flask is charged with 56 g (0.34 mole) of triethyl phosphite and heated (oil bath or mantle) to 120°. Ethyl chloroformate (34 g, 0.31 mole) is added with stirring, resulting in the immediate generation of ethyl chloride. The addition is continued at a rate so as to maintain a steady evolution of ethyl chloride. After completion of the addition, the flask is cooled, and the reaction mixture is fractionally distilled through a short column. The product has bp 122–123°/8 mm or 138–139°/12.5 mm, n_D^{20} 1.4230; the yield is 50–60%.

2. *2-Carbethoxycyclohexanone (5):* A 500-ml, three-necked, round-bottom flask is set up for distillation under aspirator pressure and is equipped with a dropping funnel (drying tube) and a magnetic stirrer. The flask is charged with 5 g (0.104 mole) of 50% suspension of sodium hydride in mineral oil and 150 ml of dry dibutyl ether. Freshly distilled ethyl diethoxyphosphinyl formate (21.2 g, 0.1 mole) is added in one portion followed by cyclohexanone (9.8 g, 0.1 mole) in several small portions. (If the stirred solution does not develop a yellow coloration, 2 to 3 drops of methanol are added.) The reaction mixture is held below 30° for 1 hour. Then, vacuum is applied, and the pot is heated to 50–60° to distil about 10 ml of the solvent together with the ethanol liberated in the reaction. The mixture is cooled to below 30° and is allowed to stand for 1 hour. It is then poured into 200 ml of anhydrous ethanol containing 15 g (approx. 0.15 mole) of concentrated sulfuric acid (the temperature rises to 50–55°). After 30 minutes, the mixture is poured into 500 ml of water and is extracted three times with 150-ml portions of benzene. The combined benzene extracts are washed with saturated sodium chloride solution and dried (anhydrous magnesium sulfate). Evaporation of the benzene (rotary evaporator) followed by distillation of the residue affords the product in about 70% yield, bp 93–94°/1 mm.

This procedure appears to be general and has been successfully applied to the following examples: ethyl acetoacetate from acetone (68%); ethyl benzoylacetate from acetophenone (74%); ethyl α-propionylpropionate from diethyl ketone (81%); ethyl 2-methylcyclohexanone-6-carboxylate from 2-methylcyclohexanone (67%).

II. Esters as Enolates: 1,4-Cyclohexanedione and Meerwein's Ester

The utility of base catalyzed condensations of esters to give β-ketoesters is well known. A straightforward example of this reaction is the intermolecular cyclization of diethyl succinate giving 2,5-dicarbethoxy-1,4-cyclohexanedione, which can in turn be easily decarboxylated to give 1,4-cyclohexanedione.

A second example of the use of esters as enolates is the formation of Meerwein's ester, an intermediate in the synthesis of substituted adamantanes (6). Dimethyl malonate and formaldehyde, on treatment with piperidine, followed by cyclization with sodium methoxide, afford the product, 1,3,5,7-tetracarbomethoxybicyclo[3.3.1]-nonane-2,6-dione, in good yield. Subsequent decarboxylation of the ester allows the isolation of bicyclo[3.3.1]nonane-2,6-dione.

A. 1,4-CYCLOHEXANEDIONE

1. *2,5-Dicarbethoxy-1,4-cyclohexanedione* (7): A 500-ml, three-necked, round-bottom flask is fitted with a condenser (drying tube) and arranged for magnetic stirring. A solution of sodium ethoxide is prepared in the flask by the addition of sodium (9.2 g, 0.4 g-atom) in small pieces to 90 ml of absolute ethanol. The mixture is heated at reflux for 3–4 hours (oil bath) to complete the reaction. Diethyl succinate (34.8 g, 0.2 mole) is added to the hot solution in one portion (*exothermic!*) and the mixture is refluxed for 24 hours. (A pink precipitate forms and persists during the reflux.)

At the conclusion of the reflux period, ethanol is removed on a rotary evaporator. Sulfuric acid (200 ml of 2 N solution) is added to the residue, and the mixture is stirred for 3–4 hours. The solid product is collected by filtration, washed with several portions of water, and allowed to dry in the air. It is then dissolved in 150 ml of boiling ethyl acetate, then filtered hot. The filtrate is cooled in ice, affording 16–17 g of 2,5-dicarbethoxy-1,4-cyclohexanedione, mp 126–129° as off-white crystals.

2. *1,4-Cyclohexanedione* (8): 2,5-Dicarbethoxy-1,4-cyclohexanedione (10 g) is suspended in a solution of 34 g of 85% phosphoric acid, 250 ml of water, and 5 ml of ethanol in a 500-ml round-bottom flask. The mixture is refluxed for 5 days (or until all the solid material has dissolved), cooled, and extracted six times with 100-ml portions of chloroform (or better, continuously extracted with chloroform). The combined chloroform extracts are dried (sodium sulfate) and the solvent is removed (rotary evaporator). The residue on distillation affords 1,4-cyclohexanedione, bp 130–133°/20 mm. The product solidifies and may be recrystallized from carbon tetrachloride, mp 77–79°.

B. MEERWEIN'S ESTER AND BICYCLO[3.3.1]NONANE-2,6-DIONE

1. CH$_3$OOCCH$_2$COOCH$_3$ + CH$_2$O $\xrightarrow[\text{2. NaOCH}_3]{\text{1. Piperidine}}$

2.

1. *Meerwein's Ester* (9): Dimethyl malonate (13.2 g, 0.4 mole) and 6 g of 40% aqueous formaldehyde solution are mixed in an Erlenmeyer flask and cooled to 0° in an ice bath. To the mixture is added 0.3 g of piperidine and enough ethanol to produce a homogeneous solution. The solution is allowed to stand at 0° for 12 hours, at room temperature for 24 hours, and at 35–40° for 48 hours. The reaction product is washed with water (50 ml) followed by dilute sulfuric acid, then dried (sodium sulfate). Unreacted malonic ester is distilled off under vacuum leaving a residue of about 12.5 g, which contains methylenemalonic ester, methylenebismalonic ester, and hexacarbomethoxypentane.

A solution of sodium methoxide is prepared by adding 1.3 g (0.056 g-atom) of sodium in small pieces to 16 ml of carefully dried methanol in a small round-bottom flask. (Alternatively, 3.1 g of dry commercial sodium methoxide can be used.) To this solution is added 10 g of the previously isolated mixture of reaction products, and the flask is heated at reflux on a steam bath for 4 hours. Methanol and methyl carbonate are then distilled insofar as possible at steam-bath temperature from the clear yellow solution. The cooled cloudy solution is then washed with ether to remove neutral materials, and the desired product is precipitated from the alkaline solution by treatment with carbon dioxide gas. The material so obtained is collected by suction filtration and washed well with water. The slightly pink crystalline powder weighs 38–40 g (56–59%). The material may be recrystallized from benzene or methanol, mp 163–164°.

2. *Bicyclo[3.3.1]nonane-2,6-dione* (10): A solution of 10 g of Meerwein's ester, 30 ml of glacial acetic acid, and 20 ml of 6 N hydrochloric acid is heated under reflux for 10 hours. The solution is then carefully distilled under aspirator pressure until all volatile materials (water and acetic acid) are removed. The solid residue is distilled at 129–131°/4 mm to afford the product. It may be recrystallized from benzene giving about 3 g of bicyclo-[3.3.1]nonane-2,6-dione, mp 138–140°.

III. Methylsulfinyl Carbanion as a Route to Methyl Ketones

On treatment with a strong base such as sodium hydride or sodium amide, dimethyl sulfoxide yields a proton to form the methylsulfinyl carbanion (dimsyl ion), a strongly basic reagent. Reaction of dimsyl ion with triphenylalkylphosphonium halides provides a convenient route to ylides (see Chapter 11, Section III), and with triphenylmethane the reagent affords a high concentration of triphenylmethyl carbanion. Of immediate interest, however, is the nucleophilic reaction of dimsyl ion with aldehydes, ketones, and particularly esters (11). The reaction of dimsyl ion with nonenolizable ketones and

$$CH_3SOCH_3 + NaH \rightarrow CH_3SOCH_2^{\ominus}Na^{\oplus} + H_2$$

$$CH_3SOCH_2^{\ominus} \xrightarrow[\text{2. } H_2O]{\text{1. } R_2C=O} R_2C-CH_2SOCH_3$$
$$\underset{OH}{|}$$

aldehydes gives good yields of β-hydroxysulfoxides. (With enolizable ketones or aldehydes, the enolization reaction is often the primary reaction.) The reaction of dimsyl ion with esters appears to be a general route to β-ketosulfoxides, providing the ester does not undergo a facile proton-transfer reaction. The importance of this reaction depends on the fact that the β-ketosulfoxide produced can be readily reduced by aluminum amalgam in water–THF to give methyl ketones. Thus, a general route from

$$RCOOR' + CH_3SOCH_2^{\ominus} \longrightarrow \left(\begin{array}{c} CH_2SOCH_3 \\ | \\ R-C-OR' \\ | \\ O^{\ominus} \end{array} \right) \longrightarrow$$

$$\underset{\underset{+ R'O^{\ominus}}{O}}{\overset{R-C-CH_2SOCH_3}{\|}} \xrightarrow{CH_3SOCH_2^{\ominus}} \underset{O}{\overset{R-C-\overset{\ominus}{C}HSOCH_3}{\|}} \xrightarrow{H_3O^{\oplus}} \underset{O}{\overset{R-C-CH_2SOCH_3}{\|}}$$

esters to methyl ketones is available.

$$\underset{O}{\overset{R-C-CH_2SOCH_3}{\|}} \xrightarrow[\text{H}_2O, \text{ THF}]{Al(Hg)} \underset{O}{\overset{R-C-CH_3}{\|}}$$

A. METHYLSULFINYL CARBANION (11)

$$CH_3SOCH_3 + NaH \xrightarrow{DMSO} CH_3SOCH_2^{\ominus}Na^{\oplus} + H_2\uparrow$$

A weighed amount of sodium hydride (50% mineral oil dispersion) is placed in a three-necked, round-bottom flask and washed three times with petroleum ether by swirling, allowing the hydride to settle, and decanting the liquid portion in order to remove the mineral oil. The flask is immediately fitted with a mechanical stirrer, a reflux condenser, and a pressure-equalizing dropping funnel. A three-way stopcock, connected to the top of the reflux condenser, is connected to a water aspirator and a source of dry nitrogen. The system is evacuated until the last traces of petroleum ether are removed from the sodium hydride and is then flushed with nitrogen by evacuating and filling with nitrogen several times. The aspirator hose is removed and this arm of the stopcock is connected to a mercury-filled U-tube, to which the system is opened. Dimethyl sulfoxide (distilled from calcium hydride, bp 64°/4 mm) is introduced through the dropping funnel and the mixture is heated with stirring to 70–75° until the evolution of hydrogen ceases. A mixture of 0.05 mole of sodium hydride (2.4 g of the 50% mineral oil dispersion) and 30 ml of dimethyl sulfoxide requires about 45 minutes for complete reaction and yields a somewhat cloudy, pale yellow-gray solution of the sodium salt, which is 1.66 M.

Powdered sodium amide reacts with dimethyl sulfoxide to generate the sodium salt under the same conditions, with the evolution of ammonia, and is comparable to sodium hydride in its reactivity.

B. REACTION OF ESTERS WITH METHYLSULFINYL CARBANION (*11*)

$$RCOOR' + 2 CH_3SOCH_2^{\ominus} \rightarrow R-\underset{\underset{O}{\|}}{C}-\overset{\ominus}{C}HSOCH_3 \xrightarrow{H_3O^{\oplus}}$$

$$R-\underset{\underset{O}{\|}}{C}-CH_2SOCH_3 \leftarrow$$

A 1.5 to 2 M solution of methylsulfinyl carbanion in dimethyl sulfoxide is prepared under nitrogen as above from sodium hydride and dry dimethyl sulfoxide. An equal volume of dry tetrahydrofuran is added and the solution is cooled in an ice bath during the addition, with stirring, of the ester (0.5 equivalent for each 1 equivalent of carbanion; neat if liquid, or dissolved in dry tetrahydrofuran if solid) over a period of several minutes. The ice bath is removed and stirring is continued for 30 minutes. The reaction mixture is then poured into three times its volume of water, acidified with aqueous hydrochloric acid to a pH of 3–4 (pH paper), and thoroughly extracted with chloroform. The combined extracts are washed three times with water, dried over anhydrous sodium sulfate, and evaporated to yield the β-ketosulfoxide as a white or pale yellow crystalline solid. The crude product is triturated with cold ether or isopropyl ether and filtered to give the product in a good state of purity.

Examples

1. ω-(Methylsulfinyl)acetophenone is prepared by the reaction of methylsulfinyl carbanion (0.07 mole) and 5.25 g (0.035 mole) of ethyl benzoate yielding approx. 5 g (78.6%) of the white solid. Recrystallization from ethyl acetate gives colorless needles, mp 86–87°.

2. Methylsulfinyl *n*-pentyl ketone is prepared by the reaction of the reagent (0.1 mole) and 7.2 g (0.05 mole) of ethyl caproate yielding the crude product as a pasty crystalline mass. The material is dissolved in hot isopropyl ether (20 ml), petroleum ether (30 ml) is added, and the solution cooled to $-15°$. Filtration yields approx. 6 g of white plates, mp 40–44°. Recrystallization from isopropyl ether affords pure product, mp 45–47°.

3. Methylsulfinyl cyclohexyl ketone is prepared by reaction of the reagent (0.08 mole) with 6.24 g (0.04 mole) of ethyl cyclohexanecarboxylate yielding about 7 g of product, mp 62–63° after recrystallization from isopropyl ether.

4. 1,11-Bis(methylsulfinyl)undecane-2,10-dione is prepared by reaction of the reagent (0.07 mole) with 3.78 g (0.0175 mole) of dimethyl azeleate yielding a pale yellow solid. Recrystallization from ethanol gives about 4 g of white product, mp 115–125°. Pure product is obtained by two additional recrystallizations from methanol–ethyl acetate, mp 129–133°.

C. REDUCTION OF β-KETO SULFOXIDES WITH ALUMINUM AMALGAM (11)

$$R-\underset{O}{\underset{\|}{C}}CH_2SOCH_3 \xrightarrow[\text{H}_2\text{O, THF}]{\text{Al(Hg)}} R-\underset{O}{\underset{\|}{C}}-CH_3$$

The compound to be reduced, dissolved in 10% aqueous tetrahydrofuran (60 ml/g of compound), is placed in a reaction vessel equipped with a stirrer. Aluminum amalgam (10 g-atom of aluminum per mole of compound) is then freshly prepared as follows. Aluminum foil is cut into strips approximately 10 cm × 1 cm and immersed, all at once, into a 2% aqueous solution of mercuric chloride for 15 seconds. The strips are rinsed with absolute alcohol and then with ether and are cut immediately with scissors (use forceps) into pieces approximately 1 cm square, directly into the reaction vessel. For the reduction of the conjugated aromatic compounds, the reaction vessel should be cooled to 0° prior to the addition of the amalgam and stirring is then continued for 10 minutes at this temperature for completion of the reduction. (Longer reaction times or higher temperatures lead to pinacol formation by further reduction of the ketone formed.) With the nonconjugated compounds, the reaction mixture is heated at 65° for 60–90 minutes after addition of the amalgam. The reaction mixture is then filtered and the filtered solids are washed with tetrahydrofuran. The filtrate is concentrated to remove most of the tetrahydrofuran, ether is added, and the ether phase is separated from the water, dried over anhydrous sodium sulfate, and evaporated to leave the ketone, usually in a high state of purity.

Examples

1. ω-(Methylsulfinyl)acetophenone gives a quantitative yield of acetophenone by the procedure, mp of the 2,4-dinitrophenylhydrazone 242–244°.
2. Methylsulfinyl n-pentyl ketone gives a quantitative yield of methyl n-pentyl ketone as a colorless liquid, mp of the 2,4-dinitrophenylhydrazone 73–74°.
3. Methylsulfinyl cyclohexyl ketone gives >95% yield of methyl cyclohexyl ketone as a colorless liquid, mp of semicarbazone 172–174°.
4. 1,11-Bis(methylsulfinyl)undecane-2,10-dione gives a quantitative yield of undecane-2,10-dione as a white solid, mp 55–58°. Recrystallization from petroleum ether raises the melting point to 61°.

IV. Cyclization with Diethyl Malonate

Intramolecular ring formation via Claisen-type condensations is usually successful only in the case of five or six membered rings. However, diethyl malonate (in the form of the enolate ion) can be employed as a nucleophile with polymethylene dihalides to carry out successful ring closure reaction yielding less favored ring sizes in satisfactory yield. The resulting 1,1-dicarboxylates can then be decarboxylated yielding cyclo-alkanecarboxylic acids. This procedure is not successful in the formation of three membered rings, but alternate procedures are readily available for the preparation of

$$Br-(CH_2)_n-Br + {}^{\ominus}CH(COOEt)_2 \rightarrow Br-(CH_2)_n-CH(COOEt)_2 \xrightarrow{\text{Base}}$$

$$Br-(CH_2)_n-\overset{\ominus}{C}(COOEt)_2 \rightarrow (CH_2)_n C(COOEt)_2$$

$$(n = 3, 4, 5, 6, 7)$$

this series (Chapter 13).

A. 1,1-CYCLOBUTANEDICARBOXYLIC ACID (12)

$$Br-(CH_2)_3-Br + EtOOC-CH_2-COOEt \xrightarrow[\text{2. KOH}]{\text{1. NaOEt}} \begin{array}{c} COOH \\ COOH \end{array}$$

A 500-ml three-necked flask equipped with a thermometer, a mechanical stirrer, a condenser, and an addition funnel (openings protected by drying tubes) is charged with 21 g (0.105 mole) of trimethylene dibromide and 16 g (0.1 mole) of diethyl malonate (previously dried over calcium sulfate). A solution of 4.6 g of sodium in 80 ml of absolute alcohol is added through the addition funnel at a rate so as to maintain the reaction temperature at 60–65°. The mixture is then allowed to stand until the temperature falls to 50–55°, then is heated on a steam bath for 2 hours. Sufficient water is added to dissolve the precipitated sodium bromide, and the excess ethanol is removed by distillation on a steam or water bath. A steam delivery tube is inserted into the flask and steam distillation is carried out until all the diethyl 1,1-cyclobutanedicarboxylate and diethyl malonate have come over (400–500 ml of distillate). The distillate is extracted three times with 50-ml portions of ether, and the ether is evaporated on a rotary evaporator (drying is not necessary). The residue is refluxed with a solution of 11.2 g of potassium hydroxide in 50 ml of 95 % ethanol for 2 hours, then cooled, and the ethanol is removed (rotary evaporator). The residue is dissolved in hot water (approx. 10 ml) and concentrated hydrochloric acid (approx. 8 ml) is added until the solution is just acid to litmus. The solution is boiled briefly to expel carbon dioxide, made slightly alkaline with dilute ammonium hydroxide, and a slight excess of aqueous barium chloride is added to the boiling solution. The solution is filtered hot to remove precipitated barium malonate, cooled, and acidified to pH 4 with concentrated hydrochloric acid (approx. 10 ml). The solution is extracted with four 35-ml portions of ether, the combined extracts are dried, and the ether is evaporated. The residual product should be air dried overnight, mp 158°, yield 5–6 g (35–42% based on the dibromide). The product may be recrystallized from ethyl acetate.

B. Cyclobutanecarboxylic Acid (12)

In a short path distilling apparatus is placed 3–5 g of 1,1-cyclohexanedicarboxylic acid. The flask is heated in an oil, sand, or metal bath to 160–170° until all the effervescence stops; then the temperature of the bath is raised to 210°. Cyclobutanecarboxylic acid distills over at 191–197°. It may be purified by redistillation at atmospheric pressure, bp 195–196°.

V. Carboxylations with Magnesium Methyl Carbonate (MMC)

When a solution of magnesium methoxide (prepared by the reaction of magnesium with methanol) is saturated with carbon dioxide, an active carboxylating agent, MMC, is produced. The reagent carboxylates substrates capable of enolization apparently by promoting formation of the magnesium chelate of the α-adduct. The reaction has been

applied successfully to nitroalkanes (13) and to ketones (14, 15), and examples of both procedures are given.

A. Magnesium Methyl Carbonate (MMC) in Dimethylformamide (13)

$$Mg + CH_3OH \rightarrow Mg(OCH_3)_2 \xrightarrow{CO_2} CH_3O\!-\!Mg\!-\!O\!-\!CO_2CH_3$$

Caution: The generation of hydrogen in this preparation necessitates the use of a hood.

A 1-liter three-necked flask is equipped with a mechanical stirrer, a condenser, and a gas inlet tube not immersed in the reaction mixture. The flask is charged with 400 ml of anhydrous methanol, stirring is begun, and 0.5 g of magnesium turnings are added. After initiation of the spontaneous reaction, an additional 23.5 g of magnesium turnings (1.0 g-atom total) is added at a rate so as to maintain a constant, controlled reflux. After the magnesium has completely reacted, the flask is heated to 50° in a water bath and excess methanol is removed under aspirator pressure while the stirring is continued. (Some methanol should be allowed to remain in the solid residue or redissolving the product will be very slow.) Heating is discontinued and dimethylformamide is added to the residue with stirring to bring the total volume to 500 ml. Carbon dioxide is then passed into the stirred solution as rapidly as it can be absorbed (rapid bubbling through a mercury-filled U-tube at the gas outlet will indicate the saturation of the system). After all the magnesium methoxide has dissolved, a fractionating column and a heating mantle are fitted on the flask. A slow stream of carbon dioxide is maintained, stirring is continued, and the flask is heated at atmospheric pressure to distil any remaining methanol. The distillation is continued until the head temperature reaches 150°. Then the mixture is cooled to room temperature and stirred under a slow carbon dioxide stream for 1 hour.

The solution is approximately 2 M in MMC and remains stable over several months.

B. α-NITROPROPIONIC ACID (13)

$$CH_3CH_2-NO_2 \xrightarrow{MMC} CH_3C{=}\overset{\oplus}{N}\overset{\overset{\ominus}{O}}{\underset{O}{\diagup}} \xrightarrow{H^\oplus} CH_3\underset{NO_2}{CHCOOH}$$

A 500-ml round-bottom flask is fitted with gas inlet and outlet tubes, a thermometer, and a magnetic stirrer. The flask is charged with 100 ml of approx. 2 M MMC solution in DMF. Nitroethane (3.8 g, 0.05 mole) is added to the flask, and the solution is heated (water bath) to 50°. The temperature is maintained with stirring for 2 hours while a slow stream of nitrogen is passed through the reaction vessel. At the end of this time, the flask is cooled to 10° in an ice bath and the contents are poured into a mixture of 80 ml of concentrated hydrochloric acid and 100 g of ice covered by a layer of 100 ml of ether. After thorough stirring, the ether layer is separated and the aqueous solution is extracted four times with 50-ml portions of ether. The combined ether extracts are dried over anhydrous magnesium sulfate and the ether is evaporated (rotary evaporator). The residue is recrystallized from chloroform–carbon disulfide giving the product in about 50% yield, mp 59–61°.

C. CARBOXYLATION OF KETONES (14)

$$R—CH_2—\overset{\underset{\|}{O}}{C}—R' \xrightarrow{\text{MMC}} R—C\overset{\|}{=}C—R' \xrightarrow{H^\oplus} R—\underset{\underset{HOOC}{|}}{C}H—\overset{\underset{\|}{O}}{C}—R'$$

The carboxylation of ketones is carried out essentially as in the preceding experiment, but at slightly higher temperatures (requiring an oil bath or mantle). Thus, acetophenone (6 g, 0.05 mole) in 100 ml of approx. 2 M MMC is stirred and heated at 110–120° for 1 hour. After cooling, hydrolysis in the acid–ice mixture, and isolation from ether, benzoylacetic acid, mp 99–100°, is obtained in 68 % yield. Similarly, 1-indanone gives 1-indanone-2-carboxylic acid, mp 100–101°, in 91 % yield.

Cyclohexanone undergoes dicarboxylation when treated with a 10-fold excess of MMC. When 2 g (0.02 mole) of cyclohexanone is treated with 100 ml of approx. 2 M MMC at 120–130° for 6 hours, the usual work-up procedure gives about 50 % yield of 2,6-cyclohexanonedicarboxylic acid, mp 123° after recrystallization from ether–petroleum ether.

VI. Alkylation of β-Ketoesters

The familiar alkylation of β-ketoesters followed by decarboxylation is still a useful route to α-alkyl ketones, although the alkylation of enamines is frequently the preferred route. Given below are two examples of alkylation of 2-carbethoxycycloalkanones (prepared in Chapter 10, Section I). In the first case, sodium ethoxide is the base employed to generate the enolate ion of 2-carbethoxycyclohexanone. In the second case, the less acidic 2-carbethoxycyclooctanone requires sodium hydride for the generation of the enolate ion.

A. DIETHYL 5-(1'-CARBOXY-2'-OXOCYCLOHEXYL)VALERATE (16)

Caution: The generation of hydrogen in this reaction requires the use of a hood.

A 500-ml, three-necked round-bottom flask is fitted with a dropping funnel and a reflux condenser. In the flask is placed 150 ml of absolute ethanol. Sodium metal (4.6 g, 0.2 g-atom) is cut into small pieces and introduced into the flask so as to maintain a rapid but controlled generation of hydrogen. When the sodium has completely dissolved, the flask is fitted with a mechanical stirrer and a heating mantle. 2-Car-bethoxycyclohexanone (34 g, 0.2 mole) is introduced into the flask and stirring is begun while the flask is brought to reflux. Ethyl 5-bromovalerate (46 g, 0.22 mole) is added to the refluxing mixture over about 30 minutes. The refluxing and stirring are continued for 12–20 hours (or until the solution is neutral to litmus). The solution is cooled and decanted from the precipitated sodium bromide. The residue is washed with two 10-ml portions of absolute ethanol and the washings are added to the main solution. Ethanol is removed by distillation from a steam bath (or by a rotary evaporator) and the residue is distilled under vacuum. The forerun contains excess ethyl 5-bromovalerate (bp 104–109°/12 mm). The product has bp 158–166°/2 mm.

B. 5-(2′-Oxocyclohexyl)valeric Acid (16)

In a 500-ml round-bottom flask fitted with a condenser, and a heating mantle is placed a mixture of 25 g of diethyl 5-(1′-carboxy-2′-oxocyclohexyl)valerate, 70 g of barium hydroxide, and 200 ml of methanol, and the mixture is refluxed for 24 hours. After cooling, the mixture is acidified (pH 4) by cautious addition of cold 10% aqueous hydrochloric acid. The acidified solution is saturated with sodium chloride and then extracted three times with 100-ml portions of chloroform. The combined chloroform extracts are dried (anhydrous magnesium sulfate) and evaporated. On vacuum distillation, the residue affords the product (about 15 g), bp 176–178°/0.5 mm.

C. 2-Methyl-2-carbethoxycyclooctanone (17)

Caution: The generation of hydrogen in this reaction requires the use of a hood.

A 500-ml, three-necked, round-bottom flask is fitted with a condenser, a magnetic stirrer, a dropping funnel, and a nitrogen inlet and outlet. All openings are protected with drying tubes. Into the flask is placed 9.6 g (0.2 mole) of 50% dispersion of sodium hydride in mineral oil. The mineral oil is removed by washing the dispersion four times with 50-ml portions of benzene, allowing the sodium hydride to settle, and removing the benzene with a pipet. Finally, 50 ml of benzene is added to the flask.

2-Carbethoxycyclooctanone (40 g, 0.2 mole) dissolved in 50 ml of dry benzene is added to the stirred sodium hydride over about 30 minutes at room temperature. The mixture is stirred an additional hour at room temperature to complete the formation of the sodium salt. Methyl iodide (284 g, 2.0 mole, a 10-fold excess) is added to the stirred solution over 1 hour and the stirring at room temperature is continued for 20–24 hours. The reaction mixture is poured cautiously into 500 ml of 3 N aqueous acetic acid, and the aqueous solution is extracted three times with 100-ml portions of benzene. The combined benzene extracts are washed three times with water and dried over anhydrous magnesium sulfate. Benzene and excess methyl iodide are removed under reduced pressure (rotary evaporator) and the residue is distilled.

D. 2-METHYLCYCLOOCTANONE (*17*)

The ketoester is mixed in a suitable round-bottom flask with excess 6 N sulfuric acid. The flask is fitted with a condenser and a mantle, and the mixture is refluxed gently for 3–4 days. The cooled reaction mixture is extracted with ether, the ether is washed with bicarbonate solution and water, then dried, and the solvent is evaporated. On distillation, the residue affords 2-methylcyclooctanone, bp 97–98°/18 mm, 86°/12 mm, n_D^{25} 1.4656.

VII. The Robinson Annelation Reaction

The Robinson annelation reaction has classically been employed for the building up of the six-membered ring components of the steroid nucleus (*18*). In the original method, the enolate is treated with the methiodide of β-diethylaminoethyl methyl ketone giving the annelated product. An improved and simplified procedure allows the Michael addition to be carried out directly on methyl vinyl ketone. The procedure gives details of the reaction on 2-methylcyclohexanone.

A. *cis*-10-METHYL-2-DECALON-9-OL (*19*)

A 250-ml, three-necked, round-bottom flask is equipped with a mechanical stirrer, a thermometer dipping in the reaction mixture, a pressure-equalizing addition funnel, a nitrogen inlet and outlet, and a Dry Ice cooling bath. In the flask is placed 56 g of 2-methylcyclohexanone, and to this is added 3 ml of 3 N ethanolic sodium ethoxide (0.7 g of sodium dissolved in 10 ml of absolute ethanol in a hood). The mixture is cooled to −10° while a nitrogen atmosphere is established and maintained. To the stirred solution, methyl vinyl ketone (35 g) is added over a 6-hour period. (The reaction mixture becomes very thick toward the conclusion of the addition.) The reaction mixture is allowed to stand an additional 6 hours at −10°, and it is then transferred with the aid of ether and saturated sodium chloride solution, as necessary, to a separatory funnel. The mixture is extracted four times with 100-ml portions of ether, and the combined ether extracts are dried over anhydrous sodium sulfate. The solution is filtered into a 2-liter Erlenmeyer flask and diluted to a total volume of 700 ml with hexane. The solution is brought to boiling on a steam bath and the ether allowed to distil off. The volume of the solution is maintained at 700 ml during the distillation by the addition of hexane as needed. When crystallization begins, the solution is allowed to cool and the crystals are collected, affording the desired product in about 40% yield, mp 120–121°. (A second crop may be obtained by reducing the volume of the filtrate, mp 104–112°. Recrystallization from ether–hexane raises the melting point.)

B. 10-METHYL-$\Delta^{1(9)}$-OCTALONE-2 (*19*)

A 7.5 g sample of *cis*-10-methyl-2-decalon-9-ol is mixed with 100 ml of 10% aqueous potassium hydroxide and steam distilled, about 1 liter of distillate being collected. The distillate is saturated with sodium chloride and extracted three times with 100-ml portions of ether. The combined ether extracts are dried (sodium sulfate) and the ether is removed (rotary evaporator). The residue is purified by vacuum distillation affording about 5 g of the colorless product, bp 82–83°/0.7 mm, 139°/15 mm.

REFERENCES

1. H. O. House, "Modern Synthetic Reactions," Chaps. 7–9. Benjamin, New York, 1965.
2. H. R. Snyder, L. A. Brooks, and S. H. Shapiro, *Org. Syn. Collective Vol.* **2**, 531 (1943).
3. A. P. Krapcho, J. Diamanti, C. Cayen, and R. Bingham, *Org. Syn.* **47**, 20 (1967).
4. A. E. Arbusov and A. A. Dunin, *J. Russ. Phys. Chem. Soc.* **46**, 295 (1914); *Chem. Abstr.* **8**, 2551 (1914); P. Nylen, *Chem. Ber.* **57**, 1023 (1924); T. Reetz, D. H. Chadwick, E. E. Hardy, and S. Kaufman, *J. Amer. Chem. Soc.* **77**, 3813 (1955).
5. I. Shahak, *Tetrahedron Lett.*, p. 2201 (1966).
6. R. C. Fort and P. von R. Schleyer, *Chem. Rev.* **64**, 277 (1964).
7. A. T. Nielsen and W. R. Carpenter, *Org. Syn.* **45**, 23 (1965).
8. W. von E. Doering and A. A. Seyigh, *J. Org. Chem.* **26**, 1365 (1961).
9. H. Meerwein and W. Schurmann, *Ann. Chem.* **398**, 196 (1913).
10. H. Stetter, H. Held, and A. Schulte-Oestrich, *Chem. Ber.* **95**, 1687 (1962).
11. E. J. Corey and M. Chaykovsky, *J. Amer. Chem. Soc.* **87**, 1345 (1965).
12. G. B. Heisig and F. H. Stodola, *Org. Syn. Collective Vol.* **3**, 213 (1955).
13. H. L. Finkbeiner and G. W. Wagner, *J. Org. Chem.* **28**, 215 (1963); H. L. Finkbeiner and M. Stiles, *J. Amer. Chem. Soc.* **85**, 616 (1963).
14. M. Stiles, *J. Amer. Chem. Soc.* **81**, 2598 (1959).
15. R. W. Griffin, J. D. Gass, M. A. Berwick, and R. S. Schulman, *J. Org. Chem.* **29**, 2109 (1964).
16. R. T. Conley and R. T. Czaja, *J. Org. Chem.* **27**, 1647 (1962).
17. S. J. Rhoads, J. C. Gilbert, A. W. Decora, T. R. Garland, R. J. Spangler, and M. J. Urbigkit, *Tetrahedron* **19**, 1625 (1963).
18. W. S. Rapson and R. Robinson, *J. Chem. Soc.*, p. 1285 (1935); W. S. Johnson, J. J. Korst, R. A. Clement, and J. Dutta, *J. Amer. Chem. Soc.* **82**, 614 (1960); E. D. Bergman, D. Ginsburg, and R. Pappo, *Org. React.* **10**, 179 (1959).
19. J. A. Marshall and W. I. Fanta, *J. Org. Chem.* **29**, 2501 (1964).

11

The Wittig Reaction

Triphenylphosphine reacts with alkyl halides to form alkyltriphenylphosphonium salts. Upon reaction with strong bases, the salts release a proton to form an ylide (alkylidenetriphenylphosphorane), which is capable of reacting with aldehydes or ketones providing an unambiguous route to olefins. Since there are virtually no

$$RCH_2X + \phi_3P \rightarrow (RCH_2\overset{\oplus}{P}\phi_3)X^{\ominus} \xrightarrow{\text{Base}} RCH{=}P\phi_3$$

$$RCH{=}P\phi_3 + R'_2C{=}O \rightarrow RCH{=}CR'_2 + \phi_3P{=}O$$

limitations on the character of the alkyl groups, the Wittig reaction has become widely employed in the synthesis of unsaturated compounds (1). The procedures given here illustrate the variety of substrates that may be used.

I. Benzyl-Containing Ylides

Phosphonium salts containing a benzyl group may be converted into ylides by the use of only moderately strong bases such as sodium ethoxide. The preparation of benzylidene derivatives of aldehydes and ketones is therefore easily done. The procedure below is for the preparation of a substituted butadiene, which in turn is ideally suited for use in the Diels-Alder reaction (see Chapter 8, Section I).

1,4-DIPHENYL-1,3-BUTADIENE (1)

$$\phi_3P + \phi CH_2Cl \rightarrow (\phi_3\overset{\oplus}{P}CH_2\phi)\overset{\ominus}{Cl} \xrightarrow{\text{NaOEt}} \phi_3P{=}CH\phi$$

$$\phi_3P{=}CH\phi + \phi CH{=}CH{-}CHO \rightarrow \phi CH{=}CH{-}CH{=}CH\phi + \phi_3P{=}O$$

The quaternary phosphonium salt is prepared by refluxing for 12 hours or longer a mixture of 4.5 g of benzyl chloride and 13 g of triphenylphosphine in 70 ml of xylene. On cooling to approx. 60°, colorless crystals of benzyltriphenylphosphonium chloride can be filtered off, washed with xylene (approx. 50 ml) and dried. The yield is virtually quantitative, mp 310–311°.

The quaternary salt is now treated with ethoxide ion in the presence of cinnamaldehyde so that the ylide reacts *in situ* as it is produced. A solution of sodium ethoxide is prepared by slowly adding 0.75 g of sodium metal to 100 ml of absolute ethanol in a dry Erlenmeyer flask (hood). In a second flask, the phosphonium chloride is dissolved in 150 ml of absolute ethanol, cinnamaldehyde (2.9 g) is added and the flask is swirled while the ethoxide solution is added; a transient orange-red color indicates the formation of the ylide.

The reaction mixture is allowed to stand at room temperature for 30 minutes, during which time crystals of the product form. The product is collected by filtration, washed with a little cold ethanol, and recrystallized from ethanol or methylcyclohexane. The faintly yellow product has mp 150–151°.

II. Alkyl Ylides Requiring *n*-Butyl Lithium

Alkyltriphenylphosphonium salts normally require extremely strong bases for the generation of the ylide. In this experiment, the alkyl group is methyl and the base required to generate the ylide is *n*-butyl lithium.

METHYLENECYCLOHEXANE (*2*)

$$\phi_3P + CH_3Br \rightarrow (\phi_3\overset{\oplus}{P}CH_3)\overset{\ominus}{Br} \xrightarrow{C_4H_9Li} \phi_3P{=}CH_2$$

$$\phi_3P{=}CH_2 \xrightarrow{\quad} {=}CH_2 + \phi_3P{=}O$$

1. *Triphenylmethylphosphonium bromide:* A pressure bottle is charged with a solution of 55 g (0.21 mole) of triphenylphosphine in 45 ml of dry benzene and cooled in an ice–salt bath. A commercially available ampoule of methyl bromide is cooled below 0° (ice–salt bath), opened, and 28 g (0.29 mole, approx. 16.2 ml) is added to the bottle in one portion. The pressure bottle is tightly stoppered, brought to room temperature, and allowed to stand for 2 days. After this time, the bottle is opened and the product is collected by suction filtration, the transfer being effected with hot benzene as needed. The yield of triphenylphosphonium bromide is about 74 g (99%), mp 232–233°. This material should be thoroughly dried (vacuum oven at 100°) before use in preparing the ylide.

2. *Methylenecyclohexane:* A 500-ml, three-necked, round-bottom flask is fitted with a mechanical stirrer, a reflux condenser leading to a mercury filled U-tube, a pressure-equalizing dropping funnel, and a gas inlet tube. The system is flushed with nitrogen

and a slight positive pressure of nitrogen is maintained thereafter. A solution of *n*-butyl lithium (0.1 mole, approx. 110 ml of the commercially available 22% solution in hexane) in 200 ml of anhydrous ether is added to the flask. Stirring is begun and triphenylphosphonium bromide (35.7 g, 0.10 mole) is added over several minutes with caution to avoid excessive frothing. The mixture is then stirred at room temperature for 4 hours. At the end of this 4-hour period, 10.8 g (0.11 mole) of redistilled cyclohexanone is added dropwise, whereupon a white precipitate forms. The mixture is then refluxed for 18–24 hours, cooled, and filtered by suction. The residue is washed with ether, and the combined filtrates are washed with several 100-ml portions of water (until neutral). The ethereal solution is dried (anhydrous magnesium sulfate) and the ether and hexane are distilled at atmospheric pressure through a good fractionating column. Fractional distillation of the residue affords methylenecyclohexane, bp 99–101°/740 mm, about 3.5 g (36%).

III. Methylsulfinyl Carbanion in the Generation of Ylides

Treatment of dimethylsulfoxide (DMSO) with sodium hydride generates methylsulfinyl carbanion (dimsyl ion), which acts as an efficient base in the production of ylides. The Wittig reaction appears to proceed more readily in the DMSO solvent, and yields are generally improved over the reaction with *n*-butyl lithium (*3*). Examples of this modification are given.

A. METHYLENETRIPHENYLPHOSPHORANE

$$CH_3SOCH_3 \xrightarrow{\text{NaH}} CH_3SOCH_2^{\ominus}Na^{\oplus} + H_2 \uparrow \xrightarrow{(\phi_3 \overset{\oplus}{P}CH_3)\overset{\ominus}{Br}}$$

$$\phi_3P{=}CH_2 + NaBr \longleftarrow$$

The general procedure for the preparation of the methylsulfinyl carbanion is described in Chapter 10, Section III. Sodium hydride (0.10 mole as a 50% dispersion in mineral oil) in a 300-ml three-necked flask is washed with several portions of petroleum ether to remove the mineral oil. The flask then is equipped with a pressure-equalizing dropping funnel, a reflux condenser fitted with a three-way stopcock, and a magnetic stirrer. The system is alternately evacuated and filled with nitrogen; 50 ml of dimethyl sulfoxide is introduced, and the mixture is heated at 75–80° for approx. 45 minutes, or until the evolution of hydrogen ceases. The resulting solution of methylsulfinyl carbanion is cooled in an ice-water bath, and 35.7 g (0.10 mole) of methyltriphenylphosphonium bromide (Chapter 11, Section II) in 100 ml of warm dimethyl sulfoxide is added. The resulting dark red solution of the ylide is stirred at room temperature for 10 minutes before use.

B. METHYLENECYCLOHEXANE (3)

$$\phi_3P{=}CH_2 + O{=}\!\!\left\langle\!\!\bigcirc\!\!\right\rangle \longrightarrow CH_2{=}\!\!\left\langle\!\!\bigcirc\!\!\right\rangle + \phi_3P{=}O$$

Freshly distilled cyclohexanone, 10.8 g (0.11 mole), is added to 0.10 mole of methylene-triphenylphosphorane, and the reaction mixture is stirred at room temperature for 30 minutes followed immediately by distillation under reduced pressure to give about 8 g (85%) of methylenecyclohexane, bp 42° (105 mm), which is collected in a cold trap.

C. 1,1-DIPHENYLPROPEN-1 (3)

$$(\overset{\oplus}{\phi_3PCH_2CH_3})\overset{\ominus}{Br} \xrightarrow{CH_3SOCH_2{}^{\ominus}Na^{\oplus}} \phi_3P{=}CHCH_3 \xrightarrow{\phi_2C{=}O}$$

$$\phi_2C{=}CHCH_3 + \phi_3PO \longleftarrow$$

A solution of sodium methylsulfinyl carbanion is prepared under nitrogen from 0.03 mole of sodium hydride and 20 ml of dimethyl sulfoxide. The solution is cooled in a cold water bath and stirred during the addition of 11.1 g (0.03 mole) of ethyltriphenyl-phosphonium bromide* in 50 ml of dimethyl sulfoxide, whereupon the deep red color of the ethylidenephosphorane is produced. After stirring at room temperature for 15 minutes, 4.55 g (0.025 mole) of benzophenone in 10 ml of dimethyl sulfoxide is added and stirring is continued for 3 hours at room temperature and $1\frac{1}{2}$ hours at 60°. The reaction mixture is cooled and poured into 200 ml of cold water in a round-bottom flask. The mixture of solid and liquid is shaken five times with 100-ml portions of pentane which are decanted, combined, washed once with water, dried over anhydrous sodium sulfate, evaporated to a volume of 50 ml, and filtered through 50 g of neutral alumina, using 1 liter of pentane to elute the product. Evaporation of the pentane gives about 4.5 g (>90%) of the product as a white crystalline solid, mp 46–49°. Recrystallization from 95% ethanol gives colorless plates, mp 49°.

IV. The Wittig Reaction Catalyzed by Ethylene Oxide

In the following procedure, use is made of the basic character of epoxides. In the presence of phosphonium salts, ethylene oxide removes hydrogen halide and the alkylidenetriphenylphosphorane is produced. If a suitable carbonyl compound is present in the reaction mixture, its reaction with the *in situ* generated phosphorane proceeds readily to give the Wittig product.

---

* This salt is prepared, by a procedure analogous to that employed for the methyl salt, from ethyl bromide and triphenylphosphine (Chapter 11, Section II).

$$(\overset{\oplus}{\phi_3 P}—CH_2 R)\overset{\ominus}{X} + CH_2—CH_2 \quad \rightleftharpoons \quad \phi_3 P{=}CH—R + HOCH_2 CH_2 X$$

$$\phi_3 P{=}CH—R + R'_2 C{=}O \quad \rightarrow \quad R'_2 C{=}CH—R + \phi_3 P{=}O$$

ETHYL CINNAMATE BY ETHYLENE OXIDE CATALYSIS (4)

$$\phi_3 P + BrCH_2 COOEt \quad \xrightarrow{\underset{CH_2—CH_2}{\overset{O}{\triangle}}} \quad \phi_3 P{=}CH—COOEt + HOCH_2 CH_2 Br$$

$$\phi_3 P{=}CH—COOEt + \phi—CHO \quad \rightarrow \quad \phi—CH{=}CH—COOEt + \phi_3 P{=}O$$

Ethylene oxide (2.5 ml, 0.05 mole) is condensed in a 50-ml round-bottom flask containing 5 ml of methylene chloride by introducing the gas via a tube into the ice-cooled flask. To the cooled flask are added triphenylphosphine (6.6 g, 0.025 mole), benzaldehyde (2.6 g, 0.025 mole), and ethyl bromoacetate (4.2 g, 0.025 mole). The flask is closed with a drying tube, brought to room temperature, and allowed to stand over-night. Fractional distillation of the solution then yields 2-bromoethanol, bp 55°/17 mm followed by the desired ethyl cinnamate, bp 142–144°/17 mm (271°/1 atm) in about 90 % yield. The residue consists of triphenylphosphine oxide, mp 150°.

V. Cyclopropylidene Derivatives via the Wittig Reaction

The reaction of cyclopropyl bromide with triphenylphosphine gives the expected phosphonium salt in less than 1 % yield. An alternate route to the salt by the thermal decomposition of 2-oxo-3-tetrahydrofuranyltriphenylphosphonium bromide gives a

virtual quantitative yield (5). Once obtained, the salt reacts normally under Wittig conditions to give cyclopropylidene derivatives in acceptable yields (6).

A. Cyclopropyltriphenylphosphonium Bromide

1. 2-Oxo-3-tetrahydrofuranyltriphenylphosphonium Bromide (7): In 100 ml of dry tetrahydrofuran are dissolved 26.2 g (0.1 mole) of triphenylphosphine and 16.5 g (0.1 mole) of α-bromo-γ-butyrolactone* (8), and the solution is refluxed for 12 hours or longer. After cooling in an ice bath, the phosphonium salt is collected by filtration and dried. It may be purified by dissolving in hot methanol (1 ml/g of salt) and precipitating by the addition of cold ethyl acetate (2.5 ml/g of salt), mp 196–197°.

2. Cyclopropyltriphenylphosphonium Bromide (5): The lactone salt is pyrolyzed by placing it in a round-bottom flask fitted with an adaptor attached to a vacuum source (aspirator is sufficient). The flask is heated (oil bath) to 180–190° for 48 hours. The residue is a virtual quantitative yield of the tan product, which may be crystallized from ethyl acetate giving cream crystals, mp 189–190°. An alternate setup is convenient if a drying pistol (Abderhalden) is available. The compound is placed in the pistol, which is then evacuated. Decalin (bp approx. 187°) is refluxed over the pistol to provide the heating source. The work-up is the same.

B. Cyclopropylidenecyclohexane (6)

The reaction is carried out in a manner similar to that described above (Chapter 11, Section II). In a 250-ml flask fitted with stirrer, condenser, and dropping funnel is placed a solution of 19.25 g (0.0505 mole) of the phosphonium salt in 180 ml of THF. The nitrogen atmosphere is established and 0.05 mole of phenyllithium added (as a solution in benzene, available from Foote Mineral Co.). The mixture is stirred for 45 minutes at room temperature and then refluxed for 15 minutes. To the red-brown solution is added dropwise over 20 minutes 4.91 g (0.05 mole) of distilled cyclohexanone; stirring is continued for 24 hours. The mixture is then concentrated by distillation at

* This compound is available from Aldrich Chemical Co.

atmospheric pressure until the benzene and THF have been removed. The residue is distilled at reduced pressure and the crude product is redistilled fractionally. The product has bp 118–119°/114 mm, obtained about 98 % pure in about 45 % yield.

C. CYCLOPROPYLIDENECYCLOPENTANE (6)

Following the procedure given above, cyclopropylidenecyclopentane is prepared in 85 % yield from 34.4 g (0.09 mole) of the phosphonium salt, 3.83 g (0.097 mole) of sodium amide (used instead of phenyllithium), and 8.4 g (0.1 mole) of cyclopentanone in ether as solvent (350 ml). The product has bp 69–70°/70 mm.

REFERENCES

1. A. Maercker, *Org. React.* **14**, 270 (1965).
2. G. Wittig and U. Schoellkopf, *Org. Syn.* **40**, 66 (1960).
3. R. Greenwald, M. Chaykovsky, and E. J. Corey, *J. Org. Chem.* **28**, 1128 (1963).
4. J. Buddrus, *Angew. Chem. Int. Ed. Engl.* **7**, 536 (1968).
5. H. J. Bestmann, H. Hartung, and I. Pils, *Angew. Chem. Int. Ed. Engl.* **4**, 957 (1965).
6. E. E. Schweizer, C. J. Berninger, and J. G. Thompson, *J. Org. Chem.* **33**, 336 (1968).
7. S. Fliszer, R. F. Hudson, and G. Salvadori, *Helv. Chim. Acta* **46**, 1580 (1963).
8. C. C. Price and J. M. Judge, *Org. Syn.* **45**, 22 (1965).

12

Reactions of Trialkylboranes

Diborane reacts with unhindered olefins to form trialkylboranes (the so-called hydroboration reaction, cf. Chapter 4). In this Chapter, several of the recently discovered carbon–carbon bond forming reactions of trialkylboranes are presented.

I. Trialkylcarbinols from Trialkylboranes and Carbon Monoxide

Carbon monoxide at atmospheric pressure reacts readily with trialkylboranes at 100–125° to give products that can be oxidized conveniently to trialkylcarbinols (*1*).

$$R_3B + CO \rightarrow (R_3CBO) \xrightarrow[\text{NaOH}]{H_2O_2} R_3COH$$

The reaction is sensitive to the presence of water, which inhibits the migration of the third alkyl group and leads to dialkyl ketones (see Chapter 12, Section II). The convenience of the hydroboration reaction combined with the use of carbon monoxide at atmospheric pressure provides the most accessible route to many trialkylcarbinols.

TRI-2-NORBORNYLCARBINOL AND OTHER TRIALKYLCARBINOLS (*1*)

A 500-ml flask is equipped with a thermometer, a magnetic stirrer, and a dropping funnel, and all openings are protected by drying tubes. The system is flushed with nitrogen and a solution of 2.84 g (0.075 mole) of sodium borohydride in 150 ml of diglyme is introduced followed by 28.3 g (0.30 mole) of norbornene. The flask is immersed in an ice-water bath and the hydroboration is achieved by the dropwise addition of 27.4 ml (0.10 mole) of boron trifluoride diglymate.* The solution is stirred

* Boron trifluoride diglymate is prepared (*2*) by mixing one part (by volume) of distilled boron trifluoride etherate with two parts (by volume) of diglyme. The ether is pumped off under vacuum (5 to 10 mm) for 20 minutes at 20°. The resulting solution is 3.65 *M* in boron trifluoride.

at room temperature for 1 hour. Ethylene glycol, 10 ml, is added, and the solution is heated and maintained at 100°. The dropping funnel is removed and replaced by a gas inlet tube connected to a cylinder of carbon monoxide. The flask is vented via an outlet tube to a mercury filled U-tube (hood). The system is flushed with carbon monoxide, and the reaction is initiated by vigorous magnetic stirring of the contents of the flask. After 1 hour, absorption is complete. The system is flushed with nitrogen and heated to 150° for 1 hour. It is then immersed in an ice-water bath and 33 ml of 6 N sodium hydroxide is added, followed by dropwise addition of 33 ml of 30% hydrogen peroxide, the temperature being maintained just under 50°. The solution is then heated to 50° for 3 hours to complete the oxidation. Addition of water (300 ml) to the cooled solution causes the precipitation of tri-2-norbornylcarbinol. The product is collected and recrystallized from pentane affording 25 g (80%) of the pure material, mp 137–138°.

Other Examples

The following trialkylcarbinols (Table 12.1) may be prepared by an analogous procedure with the time required for the absorption of carbon monoxide as shown. For liquid products, the dilution with water is followed by extraction with pentane, the pentane solution is dried, and the solvent is removed (rotary evaporator), affording the pure product.

TABLE 12.1

Olefin	Product	Time for reaction with CO (minutes)	n_D^{20}
1-Butene	Tri-n-butylcarbinol	500	1.4446
2-Butene	Tri-sec-butylcarbinol	60	1.4558
Cyclopentene	Tricyclopentylcarbinol	50	1.5128
Cyclohexene	Tricyclohexylcarbinol[a]	30	—

[a] mp 94–95°.

II. Dialkylketones from Trialkylboranes and Carbon Monoxide–Water

As mentioned in the preceding section, the presence of water during the reaction of trialkylboranes with carbon monoxide inhibits the migration of the third alkyl group and leads to production of dialkyl ketones (*3*). This fact can be employed to advantage for the preparation of dialkyl ketones as shown in the scheme.

$$R_3B + CO \xrightarrow{H_2O} \left(\begin{array}{c} RB-CR_2 \\ | \quad | \\ HO \quad OH \end{array} \right) \xrightarrow[NaOH]{H_2O_2} R_2C=O + R-OH$$

DICYCLOPENTYL KETONE AND OTHER EXAMPLES (3)

The apparatus is set up as described in the preceding experiment (Chapter 12, Section I). A solution of 20.4 g (0.30 mole) of cyclopentene in 150 ml of diglyme is introduced into the flask, which then is cooled with an ice-water bath. Hydroboration is achieved by the dropwise addition of 50 ml of a 1 M solution of diborane (0.30 mole of hydride) in tetrahydrofuran. (See Chapter 4, Section I for the external generation of diborane and the preparation of its solution in THF.) The solution is stirred at room temperature for 1 hour, and the tetrahydrofuran is removed by distillation under reduced pressure. Water (2.7 ml, 0.150 mole) is then added, and the solution is heated to 100°. The system is flushed with carbon monoxide (hood), and the reaction is initiated by rapid magnetic stirring of the contents of the flask. Absorption of the carbon monoxide begins immediately and ceases after 2½ hours. The flask is then cooled in an ice-water bath and oxidation is accomplished by the addition of 3 N sodium hydroxide followed by dropwise addition of 23 ml of 30% hydrogen peroxide, keeping the temperature under 35°. After the addition has been completed, the reaction mixture is stirred for an additional hour at 30–35°. The solution is poured into 300 ml of water and is extracted once with 100 ml of pentane. The pentane solution is back-extracted twice with 300-ml portions of water to remove diglyme, then is dried and vacuum distilled, affording 15.0 g (90%) of dicyclopentyl ketone, bp 86°/5 mm.

Other Examples

The following dialkyl ketones may be prepared by an analogous procedure with the time required for the absorption of carbon monoxide as shown (Table 12.2).

TABLE 12.2

Olefin	Product	Time for reaction with CO (minutes)	n_D^{20}
1-Butene	Di-n-butyl ketone	500	1.4201
2-Butene	Di-sec-butyl ketone	150	1.4214
1-Octene	Di-n-octyl ketone[a]	500	—
Cyclohexene	Dicyclohexyl ketone	78	1.4847
Norbornene	Di-2-norbornyl ketone[b]	150	—

[a] mp 49–50°.
[b] mp 53–54°.

III. The Reaction of Trialkylboranes with Methyl Vinyl Ketone and Acrolein

Trialkylboranes react rapidly with methyl vinyl ketone (and other α,β-unsaturated ketones) to yield, after hydrolysis, methyl ketones of the indicated structure (4). The reaction with acrolein is analogous to give β-alkylpropionaldehydes (5). The process is inefficient in that only one of the three alkyl groups of the borane is converted into product, but the rapidity and ease of carrying out the reaction may be adequate

$$R_3B + CH_2{=}CHCCH_3 \rightarrow RCH_2CH{=}\overset{CH_3}{\underset{\underset{O}{\|}}{C}}-OBR_2$$

$$\downarrow H_2O$$

$$RCH_2CH_2CCH_3 + ROH$$
$$\underset{O}{\overset{\|}{}}$$

compensation when the alkyl group is not expensive or rare.

TRIOCTYLBORANE WITH METHYL VINYL KETONE: 2-DODECANONE (4)

$$CH_3(CH_2)_5CH{=}CH_2 \xrightarrow{B_2H_6} [CH_3(CH_2)_5CH_2CH_2]_3\text{-}B$$

$$CH_2{=}CH{-}\overset{O}{\overset{\|}{C}}{-}CH_3$$

$$CH_3(CH_2)_5CH_2CH_2CH_2CH_2\overset{\|}{\underset{O}{C}}{-}CH_3$$

In a 200-ml three-necked flask fitted with a dropping funnel (drying tube) is placed a solution of 13.4 g (0.12 mole) of 1-octene in 35 ml of THF. The flask is flushed with nitrogen and 3.7 ml of a 0.5 M solution of diborane (0.012 mole of hydride) in THF is added to carry out the hydroboration. (See Chapter 4, Section I regarding preparation of diborane in THF.) After 1 hour, 1.8 ml (0.1 mole) of water is added, followed by 4.4 g (0.06 mole) of methyl vinyl ketone, and the mixture is stirred for 1 hour at room temperature. The solvent is removed, and the residue is dissolved in ether, dried, and distilled. 2-Dodecanone has bp 119°/10 mm, 245°/1 atm. (The product contains 15% of 5-methyl-2-undecane.) The reaction sequence can be applied successfully to a variety of olefins including cyclopentene, cyclohexene, and norbornene.

IV. The Reaction of Trialkylboranes with Ethyl Bromoacetate

Trialkylboranes react with ethyl bromoacetate to give ethyl alkylacetates in good yields (6). As in other reactions of boranes, only one of the three alkyl groups is utilized

$$R_3B + BrCH_2COOEt \xrightarrow{KO-t\text{-}Bu} RCH_2COOEt$$

in the product, but, again, convenience may compensate for this inefficiency.

ETHYL CYCLOPENTYLACETATE FROM CYCLOPENTENE (6)

A dry 500-ml flask equipped with a thermometer, pressure-equalizing dropping funnel, and magnetic stirrer is flushed with nitrogen and then maintained under a static pressure of the gas. The flask is charged with 50 ml of tetrahydrofuran and 13.3 ml (0.15 mole) of cyclopentene, and then is cooled in an ice bath. Conversion to tricyclo-pentylborane is achieved by dropwise addition of 25 ml of a 1 M solution of diborane (0.15 mole of hydride; see Chapter 4, Section I for preparation) in tetrahydrofuran. The solution is stirred for 1 hour at 25° and again cooled in an ice bath, and 25 ml of dry t-butyl alcohol is added, followed by 5.5 ml (0.05 mole) of ethyl bromoacetate. Potassium t-butoxide in t-butyl alcohol (50 ml of a 1 M solution) is added over a period of 10 minutes. There is an immediate precipitation of potassium bromide. The reaction mixture is filtered from the potassium bromide and distilled. Ethyl cyclopentylacetate, bp 101°/30 mm, n_D^{25} 1.4398, is obtained in about 75% yield. Similarly, the reaction can be applied to a variety of olefins including 2-butene, cyclohexene, and norbornene.

REFERENCES

1. H. C. Brown and M. W. Rathke, *J. Amer. Chem. Soc.* **89**, 2737 (1967).
2. H. C. Brown and G. Zweifel, *J. Amer. Chem. Soc.* **88**, 1433 (1966).
3. H. C. Brown and M. W. Rathke, *J. Amer. Chem. Soc.* **89**, 2739 (1967).
4. H. C. Brown et al., *J. Amer. Chem. Soc.* **89**, 5708 (1967); *J. Amer. Chem. Soc.* **90**, 4165, 4166 (1968); *J. Amer. Chem. Soc.* **92**, 710, 712 (1970).
5. H. C. Brown, M. M. Rogic, M. W. Rathke, and G. W. Kabalka, *J. Amer. Chem. Soc.* **89**, 5709 (1967).
6. H. C. Brown, M M. Rogic, M. W. Rathke, and G. W. Kabalka, *J. Amer. Chem. Soc.* **90**, 818 (1968).

13

Carbenes as Intermediates

Photolysis of ketene with light of wavelength 300–370 mμ produces the reactive intermediate carbene, which is capable of a variety of insertion and addition reactions.

$$CH_2\!=\!C\!=\!O \xrightarrow{h\nu} CH_2: \xrightarrow{CH_3CH_2CH_3} CH_3CH_2CH_2CH_3 + CH_3\underset{\underset{CH_3}{|}}{C}HCH_3$$

$\longrightarrow$ + Methylcyclohexanes

More useful for synthetic purposes, however, is the combination of the zinc–copper couple with methylene iodide to generate carbene–zinc iodide complex, which undergoes addition to double bonds exclusively to form cyclopropanes (*1*). The base-catalyzed generation of halocarbenes from haloforms (*2*) also provides a general route to 1,1-dihalocyclopropanes via carbene addition, as does the nonbasic generation of dihalocarbenes from phenyl(trihalomethyl)mercury compounds. Details of these reactions are given below.

I. Carbene Addition by the Zinc–Copper Couple

Photolytically generated carbene, as mentioned above, undergoes a variety of undiscriminated addition and insertion reactions and is therefore of limited synthetic utility. The discovery (*3*) of the generation of carbenes by the zinc–copper couple, however, makes carbene addition to double bonds synthetically useful. The iodomethylzinc iodide complex is believed to function by electrophilic addition to the double bond in a three-center transition state giving essentially cis addition. Use of the

system for the preparation of norcarane is given in the procedure.

NORCARANE (*4*)

$$\text{⬡} \quad + \quad CH_2I_2 + Zn(Cu) \quad \longrightarrow \quad \text{⬡▷} \quad + ZnI_2 + C\underset{.}{u}$$

1. *Zinc–Copper Couple:* A 500-ml Erlenmeyer flask equipped for magnetic stirring is charged with a mixture of zinc powder (49.2 g, 0.75 g-atom) and hydrochloric acid (40 ml of 3% aqueous solution). The contents of the flask are rapidly stirred for 1 minute, and the liquid is decanted. Similarly, the zinc is washed with the following: three times with 40 ml of 3% hydrochloric acid solution, five times with 100 ml of distilled water, five times with 75 ml of 2% aqueous copper sulfate solution, five times with 100 ml of distilled water, four times with 100 ml of absolute ethanol, and five times with 100 ml of absolute ether. These last ethanol and ether washes are decanted onto a Buchner funnel to prevent loss. The residue is collected by suction filtration, washed again with anhydrous ether, and dried in air. Finally, the zinc–copper couple is stored (20–24 hours) in a vacuum desiccator over phosphorous pentoxide.

2. *Norcarane:* A 500-ml round-bottom flask is equipped with a reflux condenser (drying tube), and is arranged for magnetic stirring. The flask is charged with 46.8 g (0.72 g-atom) of zinc–copper couple and 250 ml of anhydrous ether. Stirring is begun, a crystal of iodine is added, and stirring is continued until the color of the iodine is discharged. A solution of cyclohexene (53.3 g, 0.65 mole) and methylene iodide (190 g, 0.71 mole) is added rapidly to the stirred solution. Gentle refluxing (water bath) is begun and in 30–45 minutes, a mildly exothermic reaction occurs (heating discontinued, if necessary). The gentle heating is resumed and continued with stirring for 15 hours. At the conclusion of the reaction period, the ether solution is decanted from liberated copper and unreacted couple and the residue is washed with 30-ml portions of ether. The decantate and washes are combined and washed twice with 100-ml portions of saturated ammonium chloride solution (exothermic) followed by 100 ml of bicarbonate solution and 100 ml of water. The ether solution is then dried (anhydrous magnesium sulfate), filtered, and fractionated, affording norcarane, bp 116–117°, n_D^{25} 1.4546, about 35 g (56%).

II. Dibromocarbenes

Haloforms react with potassium *t*-butoxide to form dihalocarbenes, which add smoothly to olefins giving 1,1-dihalocyclopropanes (*2*). The reaction does not appear

$$HCX_3 + t\text{-BuO}^{\ominus} \rightleftharpoons {}^{\ominus}CX_3 + t\text{-BuOH}$$

$${}^{\ominus}CX_3 \rightarrow :CX_2 + X^{\ominus}$$

to be complicated by insertion and is therefore of great synthetic use. Two examples are given in the procedures.

A. 7,7-DIBROMONORCARANE (5)

In a dry, 250-ml, three-necked flask equipped with a dropping funnel and magnetic stirrer are placed 40 ml of dry *t*-butyl alcohol (distilled from calcium hydride) and 4.0 g (0.036 mole) of potassium *t*-butoxide. The solution is cooled in ice and 40 g (49 ml, 0.49 mole) of dry cyclohexene is added. Bromoform (10 g, 3.5 ml, 0.039 mole) is added to the cooled, stirred reaction vessel dropwise over about $\frac{1}{2}$ hour, and the vessel is stirred an additional hour with the ice bath removed. The reaction mixture is poured into water (approx. 150 ml), and the layers are separated. The aqueous layer is extracted with 25 ml of pentane, and the extract is combined with the organic layer. The combined layers are dried (sodium sulfate), and the solvent is removed. The product is purified by distillation, bp 100°/8 mm.

B. 9.9-DIBROMOBICYCLO[6.1.0]NONANE (6)

A dry, 500-ml, three-necked flask is fitted with a mechanical stirrer, a condenser, and a pressure-equalizing dropping funnel. The system is swept out with nitrogen and a slight positive pressure of nitrogen is maintained by venting the system to a mercury filled U-tube. In the flask is placed 21 g (0.187 mole) of dry potassium *t*-butoxide, and the flask is cooled in an ice–salt bath.

A solution of redistilled *cis*-cyclooctene (17.8 g, 0.16 mole, 21.4 ml) in 35 ml of dry pentane is added to the flask in one portion. Bromoform (42 g, 0.166 mole, 14.8 ml) is placed in the dropping funnel and added dropwise with stirring to the flask over a period of about 1 hour (color change from light yellow to brown). At the completion of the addition, the cooling bath is removed, the flask is allowed to come to room temperature, and stirring is continued at room temperature for 18–20 hours. Water (50 ml) is added, followed by sufficient hydrochloric acid (10% aqueous solution) to render the solution neutral. The organic layer is separated, and the aqueous layer is extracted three times with 15-ml portions of pentane. The combined pentane solutions are washed three times with 15-ml portions of water, then dried (anhydrous magnesium sulfate) and filtered, and the solvent is removed (rotary evaporator). The residue is distilled affording 9,9-dibromobicyclo[6.1.0]nonane, bp 62°/0.04 mm, n_D^{23} 1.5493–1.5507, about 26 g (58%).

III. Dihalocarbenes from Phenyl(trihalomethyl)mercury Compounds

Seyferth (7) discovered that phenyl(trihalomethyl)mercury compounds decompose when heated in a solvent giving dihalocarbenes. When the solvent contains a suitable olefin, carbene addition occurs giving 1,1-dihalocyclopropane derivatives. The reaction has the advantage that strong base is not required in the reaction mixture, and base-

$$\phi HgCX_3 \xrightarrow[\Delta]{\text{benzene}} \phi\text{---Hg---X} + :CX_2 \longrightarrow$$

sensitive olefins (for example, acrylonitrile) can be made to undergo carbene addition.

The preparation of the reagents, on the other hand, is somewhat involved and requires high-speed stirring (8). Given below, therefore, is an alternate method of preparation of phenyl(trichloromethyl)mercury that requires only conventional stirring (9). The use of this compound to prepare 7,7-dichloronorcarane is also given.

It should be pointed out that the mono-, di-, and tribromo derivatives of the reagent all react considerably more rapidly than the trichloro reagent. For example, the tribromo compound reacts with cyclohexene in about 2 hours, while the trichloro compound requires 36 to 48 hours (7).

A. Phenyl(trichloromethyl)mercury (9)

$$C_6H_5HgCl + Cl_3CCOONa \xrightarrow{\text{glyme}} C_6H_5HgCCl_3 + NaCl + CO_2$$

A 250-ml round-bottom flask is fitted with a reflux condenser (drying tube), a heating mantle, and a magnetic stirrer. The flask is charged with 150 ml of glyme)1,2-dimethoxyethane), 27.8 g (0.15 mole) of sodium trichloroacetate, and 31.3 g (0.1 mole) of phenylmercuric chloride. Stirring is begun, and the flask is heated to reflux, whereupon carbon dioxide evolves and sodium chloride precipitates. Refluxing is continued until the evolution of carbon dioxide ceases (about 1 hour). The mixture is cooled to room temperature, then is poured into 500 ml of water. The resulting mixture is extracted four times with 50-ml portions of ether; the combined ether extracts are washed twice with 50-ml portions of water and dried over anhydrous magnesium sulfate. Filtration followed by removal of the solvent (rotary evaporator) gives the crude product combined with phenylmercuric chloride. The solid is dissolved in 130 ml of hot chloroform and allowed to cool. The solid material thus recovered is primarily phenylmercuric chloride (mp 252°). Reduction of the solvent volume (by about 25 ml) gives a second crop of phenylmercuric chloride, and a repetition of this procedure gives a third crop (total, about 2 g). Thereafter, reductions in solvent volume followed by collection of the crystalline products yield, after several such steps, about 25 g (65%) of the desired product, mp 110°.

B. 7,7-DICHLORONORCARANE (7)

$$C_6H_5HgCCl_3 \; + \; \text{⬡} \; \longrightarrow \; \text{⬡△}<\substack{Cl \\ Cl} \; + \; C_6H_5HgCl$$

A 100-ml round-bottom flask is fitted with a reflux condenser (drying tube), a heating mantle, and a magnetic stirrer. The flask is charged with 12 g (0.03 mole) of phenyl(trichloromethyl)mercury, 7.4 g (0.09 mole) of cyclohexene, and 35 ml of benzene. The flask is flushed with nitrogen, and heating and stirring are begun. When the reflux temperature is reached, the reagents will have largely dissolved, and the slow precipitation of phenylmercuric chloride will begin. The heating and stirring are continued for 36–48 hours. The reaction mixture is now cooled and filtered. (The residue is a virtually quantitative recovery of reusable phenylmercuric chloride.) The solvent and excess cyclohexene are removed from the filtrate by fractional distillation at atmospheric pressure. Finally, the product is collected by fractional distillation in about 85% yield, bp 73–75°/10 mm, n_D^{23} 1.5018.

REFERENCES

1. L. F. Fieser and M. Fieser, "Advanced Organic Chemistry," p. 536. Reinhold, New York, 1961.
2. J. Hine, "Divalent Carbon." Ronald Press, New York, 1964; W. Kirmse, "Carbene Chemistry." Academic Press, New York, 1964.
3. H. E. Simmons and R. D. Smith, *J. Amer. Chem. Soc.* **81**, 4256 (1959).
4. R. D. Smith and H. E. Simmons, *Org. Syn.* **41**, 72 (1961).
5. W. Kemp, "Practical Organic Chemistry," p. 121. McGraw-Hill, London, 1967.
6. L. Skattebol and S. Solomon, *Org. Syn.* **49**, 35 (1969).
7. D. Seyferth, J. M. Burlitch, R. J. Minasz, J. Y. Mui, H. D. Simmons, A. J. H. Treiber, and S. R. Dowd, *J. Amer. Chem. Soc.* **87**, 4259 (1965) and references cited therein.
8. D. Seyferth and J. M. Burlitch, *J. Organometal. Chem.* **4**, 127 (1965).
9. T. J. Logan, *Org. Syn.* **46**, 98 (1966).

14

Ethynylation

Acetylenes are sufficiently acidic to react with sodium metal to generate acetylides, useful nucleophiles in the formation of carbon–carbon bonds. The reaction is classically carried out in liquid ammonia, which is a good solvent for alkali metals but which is troublesome to handle. Two convenient modifications of the acetylide generation reaction overcome this difficulty and are discussed below along with the classical method.

I. Generation of Sodium Acetylide in Liquid Ammonia

Reactions in liquid ammonia (cf. Chapter 3, Section III) require a certain amount of care, since the solvent is low boiling (−33°) and its fumes are noxious. Nevertheless, with reasonable caution, the preparation of an ammonia solution of sodium acetylide can be carried out as described. The reagent so prepared can then be directly used for displacements on alkyl halides or for additions to suitable carbonyl compounds. Examples of both reactions are given.

A. 1-HEXYNE (*1*)

$$NH_3 + Na \rightarrow NaNH_2 \xrightarrow{HC\equiv CH} NaC\equiv CH + NH_3$$

$$NaC\equiv CH + CH_3(CH_2)_3Br \rightarrow CH_3(CH_2)_3C\equiv CH + NaBr$$

Note: The following reactions must be carried out in an efficient hood.

1. *Sodamide:* A 1-liter three-necked flask is equipped with a cold-finger condenser, a mechanical stirrer, and a gas inlet tube and placed in a vessel to be used as a cooling bath. The cooling bath is filled with isopropyl alcohol or acetone, and Dry Ice is added. The cold-finger condenser should likewise be charged with a Dry Ice coolant. Liquid ammonia is run into the flask from an inclined cylinder until about 350 ml is present. A pinch of ferric nitrate (0.05 g) is added with stirring, and stirring is continued for 5–10 minutes. Sodium (0.5 g) cut in small pieces is cautiously added, and stirring is

continued until the blue color has disappeared (approx. 10 minutes). The temperature of the cooling bath is maintained below −35° by the addition of Dry Ice as necessary during the following addition of sodium. Clean sodium (13.5 g, 0.61 g-atom total employed) is added to the stirred solution in small pieces over about 1 hour. The solution becomes deep blue initially and finally turns colorless or gray. Stirring is continued until the blue color disappears; liquid ammonia is added as necessary to maintain the volume at 350 ml.

2. *Sodium Acetylide:* Commercial acetylene is purified as used by passing it through a gas trap at −80° (Dry Ice–acetone) and a mercury safety trap. The acetylene line is then attached to the gas inlet tube of the flask containing the sodium amide, and acetylene is passed into the suspension while the cooling bath is held below −35°. The addition is continued until a uniformly black liquid results (about 1 hour). Liquid ammonia is added if necessary to maintain the original volume. *Note:* If the gas inlet tube becomes clogged with solid during the course of the acetylene addition, it should be replaced.

3. *1-Hexyne:* The cooling bath is brought to approx. −50°, and the flask is fitted with a dropping funnel containing 68.5 g (53.8 ml, 0.5 mole) of *n*-butyl bromide. A slow stream of acetylene is maintained, and the bromide is added with stirring over about 1 hour. The reaction is exothermic and the bath temperature should be held at about −50°. After the addition is complete, the acetylene stream is discontinued. The cold-finger condenser and the cooling bath are filled with Dry Ice, and the mixture is allowed to stand overnight (approx. 15 hours) with slow stirring.

   With stirring, 6 g of ammonium chloride (to decompose excess acetylide) is added and the remaining ammonia is allowed to evaporate. To the residue is added cautiously 50 g of crushed ice followed by 150 ml of water; the contents are mixed and steam distilled rapidly to obtain the 1-hexyne. The organic layer is separated, dried (magnesium sulfate), and distilled through a short column. 1-Hexyne has bp 71–72°; the yield is about 28 g (68%).

B. 1-ETHYNYLCYCLOHEXANOL (2)

Employing the same procedure as described above, a solution of sodamide in 150 ml of liquid ammonia is prepared using 0.07 g of ferric nitrate and 0.2 g of sodium followed by 4.6 g of sodium (0.21 g-atom total employed). The sodamide is converted to sodium acetylide as before and to it is added with stirring a solution of 19.6 g (20.6 ml, 0.20 mole) of cyclohexanone in 30 ml of dry ether over approx. $\frac{1}{2}$ hour. Stirring is continued for 2 hours, then the sodium salt is decomposed by the gradual addition of 11.8 g (0.22 mole) of ammonium chloride. After standing overnight without cooling (by which time the ammonia will have evaporated), the residue is mixed carefully with 200 ml of cold water and the solution is extracted three times with 150-ml portions of ether. The combined ethereal extracts are washed with water, dilute sulfuric acid, and sodium bicarbonate solution, the solution is dried, and the ether is removed. The residue upon distillation affords 1-ethynylcyclohexanol, bp 83°/20 mm (mp of the pure material is 32°); the yield is about 21 g (85%).

II. The Generation of Sodium Acetylide in Tetrahydrofuran

The problems of handling liquid ammonia are alleviated in this modification of the sodium acetylide generation procedure. Finely divided sodium is prepared in boiling toluene, the toluene is replaced by THF, and a direct reaction between sodium and acetylene is carried out. The resulting sodium acetylide is employed in ethynylation reactions as before.

SODIUM ACETYLIDE IN TETRAHYDROFURAN (3)

$$Na + HC\equiv CH \xrightarrow{\text{THF}} NaC\equiv CH \xrightarrow{\text{R-Br}} R-C\equiv CH + NaBr$$

In a 250-ml three-necked flask are placed 2.3 g (0.1 g-atom) of clean sodium and 100 ml of dry toluene. The flask is fitted with a rapid mechanical stirrer and a condenser, and all openings are protected with drying tubes. The toluene is refluxed without stirring until the sodium melts, and rapid stirring is then begun until the sodium breaks into fine particles. The suspension is rapidly cooled and the toluene is removed by a siphon. The metal is washed with two 20-ml portions of dry THF and finally covered by 80 ml of the solvent. With slow stirring, a purified stream of acetylene (see preceding experiment for purification) is directed into the flask through a gas inlet tube extending below the surface of the liquid. After a brief induction period (during which the solvent becomes saturated with acetylene, 15–20 minutes) the reaction temperature increases from 20° to 45–55° in about 1 hour. The color of the mixture becomes metallic gray to white and the temperature falls. The mixture is then heated for 1 hour at 45° to complete the reaction.

Reaction with Alkyl Halides: The gas inlet tube is replaced by an addition funnel, and 10 ml of HMPT is added rapidly with stirring. The mixture is cooled to 10–15°, and a solution of the alkyl halide (0.1 mole) in 20 ml of THF is added dropwise over a period of 30–40 minutes. The mixture is then heated to 40° for 2–3 hours. The thick white suspension of the sodium halide is cooled and dilute cold hydrochloric acid is carefully added until the mixture is clear. The organic layer is separated, and the aqueous layer is extracted three times with 20-ml portions of ether, the ethereal extracts then being combined with the organic material. The ethereal solution is washed twice with saturated sodium chloride solution and dried. The ether and THF are removed under reduced pressure (rotary evaporator), and the alkyne is distilled.

Examples

1. 1-Heptyne, bp 99–100°, n_D^{20} 1.4084, is obtained in 57% yield from 1-bromopentane.
2. 1-Nonyne, bp 55°/23 mm, 150–151°/1 atm, n_D^{20} 1.4221, is obtained in 72% yield from 1-bromoheptane.
3. 1,9-Decadiyne, bp 71°/14 mm, 76–77°/18 mm, n_D^{20} 1.4532, is obtained in 71% yield from 1,6-dibromohexane (hexamethylene dibromide).

III. The Generation of Sodium Acetylide via Dimsylsodium

The most convenient method for the preparation of sodium acetylide appears to be by reaction of acetylene with sodium methylsulfinyl carbanion (dimsylsodium). The anion is readily generated by treatment of DMSO with sodium hydride, and the direct introduction of acetylene leads to the reagent. As above, the acetylide may then be employed in the ethynylation reaction.

SODIUM ACETYLIDE IN DIMETHYL SULFOXIDE (*4*)

$$CH_3SOCH_3 \xrightarrow[\text{DMSO}]{\text{NaH}} CH_3SO_2CH_2^{\ominus}Na^{\oplus} \xrightarrow{\text{HC}\equiv\text{CH}} NaC\equiv CH$$

Methylsulfinyl carbanion (dimsyl ion) is prepared from 0.10 mole of sodium hydride in 50 ml of dimethyl sulfoxide under a nitrogen atmosphere as described in Chapter 10, Section III. The solution is diluted by the addition of 50 ml of dry THF and a small amount (1–10 mg) of triphenylmethane is added to act as an indicator. (The red color produced by triphenylmethyl carbanion is discharged when the dimsylsodium is consumed.) Acetylene (purified as described in Chapter 14, Section I) is introduced into the system with stirring through a gas inlet tube until the formation of sodium acetylide is complete, as indicated by disappearance of the red color. The gas inlet tube is replaced by a dropping funnel and a solution of 0.10 mole of the substrate in 20 ml of dry THF is added with stirring at room temperature over a period of about 1 hour. In the case of ethynylation of carbonyl compounds (given below), the solution is then cautiously treated with 6 g (0.11 mole) of ammonium chloride. The reaction mixture is then diluted with 500 ml of water, and the aqueous solution is extracted three times with 150-ml portions of ether. The ether solution is dried (sodium sulfate), the ether is removed (rotary evaporator), and the residue is fractionally distilled under reduced pressure to yield the ethynyl alcohol.

Examples

1. 1-Butyn-3-ol is produced in 78% yield by the reaction of acetaldehyde with the reagent. The product has bp 106–108°/1 atm, n_D^{20} 1.4265.
2. 1-Ethynylcyclohexanol (see Chapter 14, Section I) may be prepared by the modification, bp 83°/20 mm.

REFERENCES

1. K. N. Campbell and B. K. Campbell, *Org. Syn. Collective Vol.* **4**, 117 (1963); A. I. Vogel, "Practical Organic Chemistry," 3rd ed., p. 901. Longmans, London, 1956.
2. J. H. Saunders, *Org. Syn. Collective Vol.* **3**, 416 (1955).
3. J.-F. Normant, *Bull. Soc. Chim. Fr.*, p. 859 (1965).
4. J. Krig, M. J. Benes, and J. Peska, *Tetrahedron Lett.*, p. 2881 (1965).

15

Structural Isomerizations

The following reactions share only the feature that they are structural isomerizations. They incorporate a variety of reagents and procedures from acid catalysts to photolysis.

I. Acid Catalyzed Rearrangement of Saturated Hydrocarbons

The rearrangement of certain saturated hydrocarbons under the influence of Lewis acids has been known for some time. Cyclohexane, for example, is converted to a mixture of hydrocarbons including methylcyclopentane. It was while investigating this

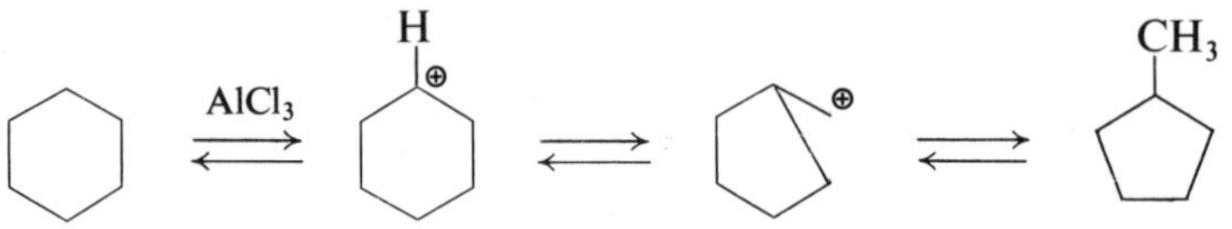

and related phenomena that Schleyer (*1*) discovered that tetrahydrodicyclopentadiene upon treatment with aluminum chloride produced adamantane, a remarkable hydrocarbon available previously only in low concentration in certain European crude petroleums or by a laborious many-step synthesis. Below is given Schleyer's synthesis as well as a modification wherein the product is isolated by a thiourea inclusion complex.

A. Adamantane (*2*)

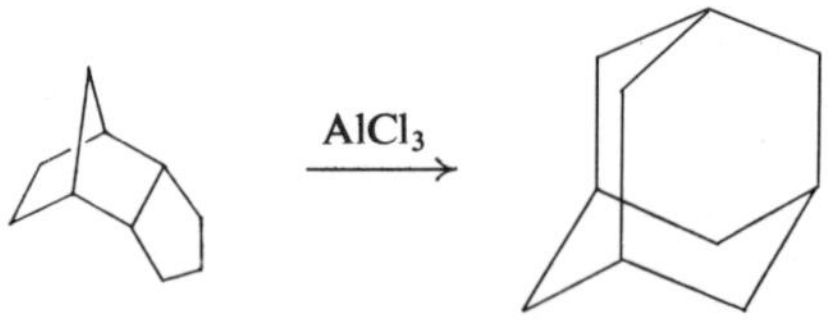

A 125-ml standard taper Erlenmeyer flask is equipped for magnetic stirring and charged with 50 g (0.37 mole) of tetrahydrodicyclopentadiene (Chapter 5, Section I). A standard taper inner joint or a short condenser is greased and inserted into the flask to serve as an air condenser. Anhydrous aluminum chloride (10 g) is added in one portion and the stirred mixture is heated to 150–180° (oil bath or stirrer-hot plate). Initially, aluminum chloride sublimes into the condenser; it should be reintroduced into the flask with a spatula. The heating and stirring are continued for 8–12 hours, then the flask is cooled whereupon two layers form. The upper layer (containing the adamantane) is separated by decantation and the residue is washed five times with 15-ml portions of 30–60° petroleum ether. The combined decantate and washings are warmed until the adamantane is in solution. Decolorization of the warm solution is then carried out by the careful addition of 2.5–3.0 g of chromatography grade alumina, filtration of the hot solution, and thorough washing of the residue with petroleum ether. Concentration of the filtrate (steam bath) to 50 ml followed by cooling in a Dry Ice–acetone bath affords the crystalline product, which is collected by suction filtration. The yield of adamantane, mp 255–260° (sealed capillary), is about 7 g (14%). Recrystallization from petroleum ether affords material of mp 268–270°.

B. Modified Isolation Procedure (3)

After treatment with chromatography-grade alumina followed by filtration, the petroleum ether solution of the product is poured into a solution of 50 g of thiourea in 750 ml of methanol with stirring of the two layers to allow complete formation of the inclusion complex (about 10 minutes). The complex is collected by suction filtration and washed with petroleum ether. (The dried complex weighs 30–35 g.) The complex (drying unnecessary) is decomposed by vigorous stirring in a mixture of 400 ml of water and 200 ml of ether. The ethereal layer is separated and dried (magnesium sulfate) and the solvent is evaporated, affording the product, mp 258–265°, in 29–32% yield. Recrystallization from isopropyl alcohol (13 ml/g) gives material of mp 268–270° (sealed capillary).

II. Photolytic Ring Contraction

In his investigations of strained bicyclic systems, Meinwald (4) has explored the photolytic rearrangement of α-diazoketones to carboxylic acids according to the reactions. The procedure given below is an example of this process, employing as the starting material camphor quinone (Chapter 1, Section XI) and resulting in the formation of Horner's acid (5).

1,6,6-Trimethylbicyclo[2.2.1]hexane-5-carboxylic Acid (6)

1. *Camphor Quinone Monotosylhydrazone:* To 22 g (0.133 mole) of camphor quinone in 150 ml of chloroform is added 26.1 g (0.14 mole) of *p*-toluenesulfonylhydrazide in one batch. The reaction vessel is protected with a drying tube and fitted with a magnetic stirrer, and the mixture is stirred at room temperature for 24 hours. The monotosylhydrazone is not isolated.

2. *Diazocamphor:* A large wide-diameter chromatography column (8.5 cm diameter) is packed with 1000 g of basic chromatography grade alumina. The previously prepared solution of the monotosylhydrazone is filtered, if necessary, to remove solids, then poured directly onto the alumina, and the column is eluted with chloroform. The solution of diazocamphor thus obtained is evaporated yielding the desired product. It may be recrystallized from hexane, mp 75°.

3. *1,6,6-Trimethylbicyclo[2.1.1]hexane-5-carboxylic Acid:* A solution of 17.8 g (0.10 mole) of diazocamphor in 500 ml of dioxane and 400 ml of water is deoxygenated with a slow stream of nitrogen for 20 minutes. The solution is then irradiated for $3\frac{1}{2}$ hours in a quartz vessel using a 500 watt Hanovia mercury lamp. (The completion of the reaction is verified by the disappearance of the diazo IR absorption at 4.85 μ.) The solution is placed on a rotary evaporator to remove about $\frac{3}{4}$ of the solvent, and the remaining aqueous solution is made definitely alkaline by the addition of aqueous sodium carbonate. The alkaline solution is washed several times with ether to remove neutral materials, and the ether is discarded. The aqueous solution is then carefully acidified with cold dilute sulfuric acid to pH 4. The mixture is extracted twice with ether, and the combined ether extracts are washed with water and dried. Evaporation of the ether gives the crude product, which may be recrystallized from aqueous methanol, mp 100–101°.

III. Isomerization of 1-Ethynylcyclohexanol: Three Methods

The readily prepared 1-ethynylcyclohexanol (Chapter 14) can be isomerized with acidic catalysts into 1-acetylcyclohexene, a compound of considerable use in the building up of polycyclic molecules. To illustrate the fact that a variety of conditions can be employed to carry out a given process, three methods are presented.

1-ACETYLCYCLOHEXENE

1. *Catalysis by Phosphorous Pentoxide (7):* A 500-ml round-bottom flask is charged with a mixture of 1-ethynylcyclohexanol (40 g, 0.32 mole), 250 ml of dry benzene, and 10 g of phosphorous pentoxide. (The addition of the phosphorous pentoxide may be attended by considerable heating if the benzene is not well dried; no particular disadvantage is found in this case, providing provision for initial cooling is made.) A condenser is attached to the flask, and the contents are refluxed gently (steam bath) for $2\frac{1}{2}$ hours. The cooled solution is then decanted from the phosphorous pentoxide, washed once with bicarbonate solution, and dried (anhydrous sodium sulfate). Removal of the benzene (rotary evaporator) and fractionation of the residue affords the desired product, bp 85–88°/22 mm, n_D^{20} 1.4892, about 25 g (61 %).

2. *Catalysis by an Acidic Resin (8):* The resin (Dowex-50, 200–400 mesh or comparable sulfonated polystyrene) is prepared for use by suspending it in dilute sulfuric acid, followed by many washings with water (and decantation of the wash), and air drying for several days.

A stirred mixture of 39.0 g of 1-ethynylcyclohexanol, 100 ml of acetic acid, 10 ml of water and 20 g of resin is heated at reflux for 45 minutes. As soon as reflux temperature is reached the resin begins to darken and in a few minutes changes from the original brown to almost black. After cooling, the resin is filtered off and washed with ether. The filtrate and washings are diluted with water and made slightly alkaline by cautious addition of 40% sodium hydroxide solution. The organic product is taken into ether and the ether solution washed with saturated sodium chloride solution and filtered through anhydrous magnesium sulfate. The ether is removed under vacuum and the residue is distilled as above. The yield is about 85%, n_D^{26} 1.4872.

3. *Catalysis by Formic Acid (9).* A mixture of 65 g (0.5 mole) of 1-ethynylcyclohexanol and 400 ml of formic acid (90%) is gently heated under reflux until a vigorous reaction ensues. After heating under reflux for 45 minutes the mixture is poured into 2 liters of

ice water and then extracted with pentane. The pentane solution is washed with sodium hydroxide solution (10%), and the pentane is removed under vacuum. Steam distillation of the residue provides the crude product, which is separated from water, dried, and distilled at reduced pressure as above. The yield is about 32 g (49%), n_D^{20} 1.4900.

IV. Photolytic Isomerization of 1,5-Cyclooctadiene

Photolytic reactions of dienes frequently give complex mixtures of rearranged products. Described here, however, is a photolytic isomerization of 1,5-cyclooctadiene (present in solution, in part, as a complex with cuprous chloride) that affords a good yield of one product.

TRICYCLO[3.3.0.0^{2,6}]OCTANE (10)

A 2-liter Vycor flask is equipped with a condenser and a drying tube and surrounded by a bank of sixteen GE 68T5 lamps. In the flask is placed a solution of 25.3 g (0.234 mole) of 1,5-cyclooctadiene and 0.1 g of cuprous chloride in 1.8 liter of ether. The flask is then irradiated for 20 hours. The reaction mixture is filtered to remove a dark green precipitate, and the walls of the flask are washed with nitric acid. The filtrate is returned to the flask and the irradiation is continued for an additional 4 days, whereupon the filtration and washing procedure is repeated, and the irradiation resumed. After the tenth day of irradiation, the filtration and washing procedure is again repeated and fresh cuprous chloride (0.05 g) is added. The reaction is stopped after 15 days of irradiation, and the mixture is carefully fractionated at atmospheric pressure. The fraction boiling at 120–126° (12–15 g of about 90% product) is further purified by refluxing with a solution of potassium permanganate (20 g) in 200 ml of water for 1 hour followed by steam distillation of the mixture. The distillate is extracted with pentane, the pentane solution is dried, and the solution is fractionally distilled giving about 11 g of tricyclo[3.3.0.0^{2,6}]octane.

V. Oxidative Rearrangement of β-Diketones

Certain β-diketones react in the presence of hydrogen peroxide to give rearranged carboxylic acids. A proposed (11) mechanism is shown. In the procedure, the reaction is applied to 2-acetylcyclohexanone (Chapter 9, Section II).

CYCLOPENTANECARBOXYLIC ACID (*11*)

A 250-ml flask is charged with 28 g (0.20 mole) of 2-acetylcyclohexanone and 25 g (0.22 mole) of 30% hydrogen peroxide in 100 ml of *t*-butyl alcohol. The solution is refluxed for 3 hours, cooled, and a pinch of palladium on charcoal (10%) is cautiously added. The mixture is refluxed for an additional $\frac{1}{2}$ hour to decompose excess peroxide. The cooled mixture is then filtered through celite, and the volume is reduced by removal of *t*-butyl alcohol and water at reduced pressure. Distillation of the residue affords about 85% of cyclopentanecarboxylic acid, bp 59–62°/1 mm, 123°/27 mm, 215–216°/1 atm.

VI. Base Catalyzed Rearrangement of 4-Benzoyloxycyclohexanone

An intriguing rearrangement was reported by Yates (*12*) during an attempt to carry out the Stobbe condensation. It resulted in the formation of a cyclopropane derivative from 4-benzoyloxycyclohexanone by the (proposed) mechanism shown.

A similar experiment on 4-benzoyloxycycloheptanone gave parallel results. The procedure gives the details of the former experiment. (The preparation of 4-benzoyloxycyclohexanone is described in Chapter 7, Section X).

2-BENZOYLCYCLOPROPANEPROPIONIC ACID (*12*)

t-Butyl alcohol is dried by refluxing it for 3 hours with calcium hydride (about 3 g/100 ml), followed by distillation at atmospheric pressure.

A 250-ml three-necked flask is fitted with a condenser (drying tube). The system is flushed with dry nitrogen, and a dry nitrogen atmosphere is maintained. In the flask is placed a solution of potassium *t*-butoxide (2.8 g, 0.025 mole) in dry *t*-butyl alcohol (100 ml). 4-Benzoyloxycyclohexanone (5 g, 0.022 mole, Chapter 7, Section X) is added to the solution, the transfer being assisted by the use of 10–15 ml of dry *t*-butyl alcohol. The mixture is cautiously brought to reflux, and refluxing is continued for 45 minutes. The mixture is then cooled rapidly to room temperature and carefully acidified by the addition of 10 ml of 6 *N* hydrochloric acid (potassium chloride will precipitate). The mixture is placed on a rotary evaporator and the bulk of the solvent is removed. The residue is diluted with sufficient water to dissolve the potassium chloride and extracted three times with 50-ml portions of ether. The ether extracts are combined and extracted four times with 100-ml portions of aqueous 5% sodium bicarbonate solution. The bicarbonate extracts are combined and the solution is acidified by the addition of concentrated hydrochloric acid to pH 4. The mixture is now extracted three times with 100-ml portions of ether, the combined ethereal extracts are washed with water, then dried, and the solvent is removed. The residual product may be recrystallized from benzene–hexane. The acid has mp 65–68°.

VII. Allenes from 1,1-Dihalocyclopropanes by Methyllithium

The treatment of 1,1-dihalocyclopropanes by methyllithium appears to be a general route to allenes, providing the product is not highly strained. An example is shown in the reaction (*13*). The procedure given below employs the reaction for the preparation

of 1,2-cyclononadiene in good yield. (For preparation of the starting material, see Chapter 13, Section II.)

1,2-CYCLONONADIENE (*14*)

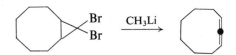

A 250-ml, three-necked, round-bottom flask is equipped with a mechanical stirrer, a pressure-equalizing dropping funnel, and a nitrogen inlet. The flask is charged with a solution of 18.7 g (11.6 ml, 0.66 mole) of 9,9-dibromobicyclo[6.1.0]nonane (Chapter 13, Section II) in 15 ml of anhydrous ether, while in the dropping funnel is placed an ethereal solution of methyllithium (0.085 mole, 42.5 ml of the commercial 2 M solution). The flask is cooled and maintained at $-30°$ to $-40°$ (Dry Ice–acetone), stirring is begun, and the ethereal methyllithium is added dropwise over about 20 minutes. Stirring is continued for 30 minutes in the cold, and for an additional 30 minutes as the flask is allowed to come to room temperature. Water (approx. 10 ml) is added dropwise to decompose the excess methyllithium, followed by an additional 40 ml of water. The ether layer is separated, and the aqueous phase is extracted three times with 15-ml portions of ether. The combined ether layer and extracts are washed with 10-ml portions of water (until neutral) and dried (anhydrous magnesium sulfate). The ether is removed by distillation through a short column at atmospheric pressure. Fractional distillation of the residue affords 1,2-cyclononadiene, bp 62–63°/16 mm, n_D^{20} 1.5060, about 6 g (80%).

REFERENCES

1. P. von R. Schleyer, *Tetrahedron Lett.*, p. 305 (1961).
2. P. von R. Schleyer, M. M. Donaldson, R. D. Nicholas, and C. Cupas, *Org. Syn.* **42**, 8 (1962).
3. A. Ault and R. Kopet, *J. Chem. Educ.* **46**, 612 (1969).
4. J. Meinwald, *Rec. Chem. Progr.* **22**, 39 (1961).
5. L. Horner and E. Spietschka, *Chem. Ber.* **88**, 934 (1955).
6. J. Meinwald and J. K. Crandall, *J. Amer. Chem. Soc.* **88**, 1292 (1966); R. R. Pennelly and J. C. Shelton, personal communication.
7. J. H. Saunders, *Org. Syn. Collective Vol.* **3**, 22 (1955).
8. M. S. Newman, *J. Amer. Chem. Soc.* **75**, 4740 (1953).
9. J. D. Chanley, *J. Amer. Chem. Soc.* **70**, 244 (1948).
10. R. Srinivasan, *J. Amer. Chem. Soc.* **85**, 3048 (1963); J. Meinwald and B. E. Kaplan, *J. Amer. Chem. Soc.* **89**, 2611 (1967).
11. G. B. Payne, *J. Org. Chem.* **26**, 4793 (1961).
12. P. Yates and C. D. Anderson, *J. Amer. Chem. Soc.* **85**, 2937 (1963).
13. L. Skattebol, *Chem. Ind.* (*London*), p. 2146 (1962); *J. Org. Chem.* **31**, 2789 (1966).
14. L. Skattebol and S. Solomon, *Org. Syn.* **49**, 35 (1969).

16

Elimination, Substitution, and Addition Reactions Resulting in Carbon–Carbon Bond Formation

The following experimental procedures do not fall into any convenient categories, but all are of interest in that they require reagents and techniques of general importance.

I. Carboxylation of Carbonium Ions

In the presence of strong acid, formic acid decomposes to water and carbon monoxide. In the process, reactive intermediates form which are capable of direct carboxylation of carbonium ions. Since many carbonium ions are readily generated by the reaction of alcohols with strong acid, the process of elimination and carboxylation can be conveniently carried out in a single flask. The carbonium ions generated are subject to the

$$R-OH \xrightarrow{\text{H}_2\text{SO}_4} (R^{\oplus}) \xrightarrow[\text{H}^{\oplus}]{\text{HCOOH}} R-COOH$$

usual rearrangements, and mixtures of product carboxylic acids may result if two or more carbonium ions of comparable stability are formed. t-Butyl alcohol, for example, gives only trimethylacetic acid (in 75% yield) under the reaction conditions, whereas primary or secondary alcohols usually give mixtures (1). It is nevertheless possible to exercise some control over the distribution of products by varying reaction conditions as is illustrated by the procedure for the preparation of the isomeric 9-decalincarboxylic acids. This and another example are detailed.

A. 1-METHYLCYCLOHEXANECARBOXYLIC ACID (1)

Caution: Because carbon monoxide is evolved, the reaction should be carried out in a good hood.

A 1-liter, three-necked, round-bottom flask is fitted with a mechanical stirrer, a pressure-equalizing dropping funnel, and a thermometer immersed in the reaction mixture. The flask is charged with 270 ml (497 g, 4.86 mole) of 96% sulfuric acid and cooled to 15–20° with rapid stirring. This temperature is maintained throughout the addition. Formic acid (3 ml of 98–100%) is added to the flask and stirred until the evolution of carbon monoxide becomes copious (about 5 minutes). Then, a mixture of 2-methylcyclohexanol (28.5 g, 0.25 mole) in 98–100% formic acid (46 g, 1.0 mole) is added over a period of about 1 hour. After an additional hour's stirring at 15–20°, the reaction mixture is poured with stirring into a 4-liter beaker containing 1 kg of ice, whereupon the crude product separates as a white solid. The mixture is extracted three times with 200-ml portions of hexane, and the combined hexane extracts are in turn extracted twice, each time with a mixture of 175 ml of 1.4 N potassium hydroxide solution and 50 g of ice. The combined alkaline extracts are washed with 100 ml of hexane (discard), and acidified (about 35 ml of concentrated hydrochloric acid) to pH 2. The acidic aqueous suspension of the product is extracted three times with 100-ml portions of hexane; the combined hexane extracts are washed with 75 ml of water and dried (anhydrous sodium sulfate). Removal of the hexane (rotary evaporator) leaves a residue of 1-methylcyclohexanecarboxylic acid, mp 34–36°, about 35 g (98%). Distillation, if desired, gives about 32 g (92%) of the purified acid, bp 132–140°/19 mm, 79–81°/0.5 mm, mp 38–39°.

B. *cis*-9-DECALINCARBOXYLIC ACID (2)

The apparatus described in the preceding experiment is employed. To 415 g of 98% sulfuric acid is added 200 g of 30% fuming sulfuric acid and the solution cooled to approx. 10°. About 3 ml of 88% formic acid is added and the rapidly stirred solution cooled to 5°. After 5 minutes, 46 g of 88% formic acid and 50 g of decahydro-2-naphthol are added from two dropping funnels over ½ hour. The solution should foam greatly during the additions. After stirring at approx. 0° for ½ hour longer, the solution is poured on ice. The oil, which soon crystallizes, is dissolved in ether and extracted into 10% sodium carbonate solution. Acidification of the aqueous layer gives about 80% of 9-decalincarboxylic acid which is largely cis. After three recrystallizations from acetone, the pure product is obtained, mp 122–123° (yield about 7 g).

C. *trans*-9-Decalincarboxylic Acid (*2*)

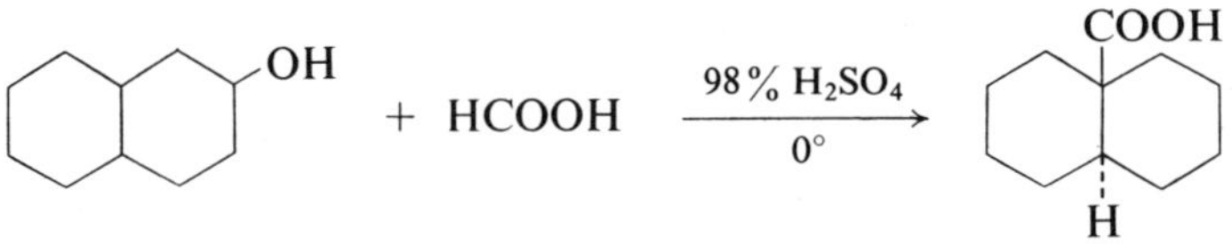

$$\text{(decalinol)} + \text{HCOOH} \xrightarrow[0°]{98\% \text{ H}_2\text{SO}_4} \text{(decalincarboxylic acid)}$$

The apparatus employed in the preceding experiment is used. To 600 g of 98% sulfuric acid at 0° (ice–salt bath) is added about 3 ml of 88% formic acid. When the rapidly stirred solution becomes foamy with evolution of carbon monoxide, 50 g of decahydro-2-naphthol and 100 g of 88% formic acid are added from two dropping funnels over 3 hours. During the addition, the temperature is kept below 5°; the mixture continues to foam. Work-up as for the cis acid gives about 85% of solid acid, predominantly trans. After three recrystallizations from acetone, about 7 g of the pure acid is obtained, mp 135–136°.

II. Paracyclophane via a 1,6-Hofmann Elimination

The Hofmann elimination is a classic route to olefins via alkylammonium hydroxides from amines. In the present instance it is employed in the generation of *p*-xylylene, a

$$\text{R}_2\text{CH—CR}_2\text{—NH}_2 \xrightarrow[\text{2. Ag}_2\text{O}]{\text{1. CH}_3\text{I}} \text{R}_2\text{CH—CR}_2\text{—N}^{\oplus}(\text{CH}_3)_3\text{OH}^{\ominus}$$

$$\downarrow \Delta$$

$$\text{R}_2\text{C}{=}\text{CR}_2 + (\text{CH}_3)_3\text{N} + \text{H}_2\text{O}$$

highly reactive intermediate that undergoes rapid dimerization and polymerization giving as the dimeric paracyclophane, a compound of some theoretical interest.

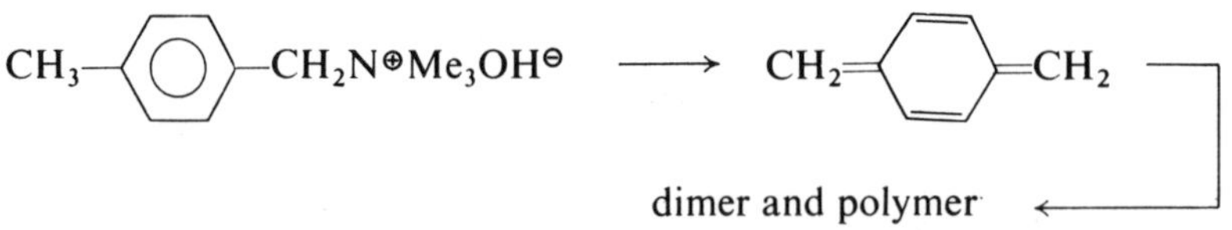

dimer and polymer

[2.2]Paracyclophane (*3*)

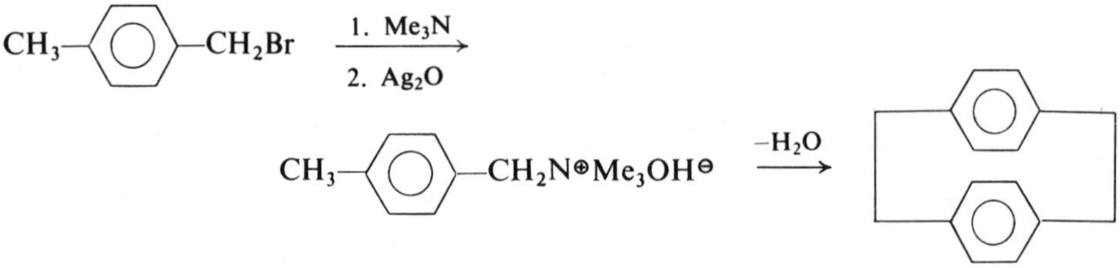

Caution: This preparation should be conducted in a hood to avoid exposure to trimethylamine and to α-bromo-p-xylene.

1. *p-Methylbenzyltrimethylammonium Bromide:* A 500-ml, three-necked, round-bottom flask is fitted with a mechanical stirrer, a condenser (drying tube), and a gas inlet tube placed just above the liquid surface. The flask is charged with a mixture of 300 ml of anhydrous ether and 50 g (0.27 mole) of α-bromo-p-xylene, and cooled with stirring in an ice bath. An ampoule of anhydrous trimethylamine is cooled in an ice–salt bath and opened, and 25 g (0.43 mole) is transferred to a small Erlenmeyer flask containing a boiling chip and fitted with a gas outlet tube leading to the above gas inlet tube. By controlled warming, the amine is distilled into the stirred reaction vessel over a period of about 1 hour (separation of solid). The resulting mixture is allowed to stand overnight at room temperature, the product is collected by suction filtration, washed with 100 ml of anhydrous ether, and air dried, affording about 65 g (99%) of p-methylbenzyltrimethylammonium bromide.

2. *[2.2]Paracyclophane:* An aqueous solution of p-methylbenzyltrimethylammonium hydroxide is prepared by dissolving p-methylbenzyltrimethylammonium bromide (24 g, 0.10 mole) in 75 ml of water, adding silver oxide* (23 g) in one portion, and stirring the mixture (magnetic stirrer) at room temperature for 1.5 hour. The mixture is then filtered and the residue is washed with 40 ml of water. The combined filtrate and wash (approx. 110 ml total) are placed in a 500-ml, three-necked, round-bottom flask fitted with a mechanical stirrer and a water separator surmounted by a condenser. Toluene (300 ml) and 0.5 g of phenothiazine (a polymerization inhibitor) are added to the flask, stirring is begun, and the flask is maintained at reflux (mantle) for about 3 hours. During this period, the water will be deposited in the trap and should be drained periodically. When the pot mixture is free of water, decomposition of the tetraalkylammonium hydroxide begins, trimethylamine is liberated, and solid poly-p-xylylene separates. The decomposition is complete after about $1\frac{1}{4}$ hour (cessation of trimethylamine evolution). The mixture is cooled and the solid is collected by suction filtration (*filtrate reserved*). The solid is then extracted (18–24 hours) in a Soxhlet apparatus employing as solvent the above filtrate. After the completion of the extraction, the toluene solution is evaporated (rotary evaporator) and the solid residue washed three times with 10-ml portions of acetone. Final purification of the product is carried out by sublimation at 0.5–1.0 mm with an oil bath temperature of 150–160°. The sublimate is [2.2]paracyclophane as white crystals, mp 284–287° (sealed capillary), about 1.0 g (10%).

III. Diphenylcyclopropenone from Commercial Dibenzyl Ketone

Cyclopropanes undergo a ready reaction with bromine to give 1,3-dibromopropane, and can in turn be formed from 1,3-dibromopropane by the Wurtz reaction. A variation

---

* Commercially available silver oxide is satisfactory.

on this scheme is employed in the procedure to convert α,α'-dibromodibenzyl ketone to the corresponding cyclopropenone.

DIPHENYLCYCLOPROPENONE (4)

$$C_6H_5CH_2CCH_2C_6H_5 + Br_2 \rightarrow C_6H_5CHCCHC_6H_5 \xrightarrow{Et_3N}$$

1. *α,α'-Dibromodibenzyl Ketone:* A 500-ml one-necked flask is fitted with a magnetic stirrer and a pressure-equalizing addition funnel. The flask is charged with a solution of dibenzyl ketone (17.5 g, 0.083 mole) in 65 ml of glacial acetic acid. Stirring at room temperature is begun and a solution of 27.5 g (0.167 mole) of bromine in 125 ml of acetic acid is added over 15 minutes. The solution is stirred for an additional 5 minutes and is then poured into 250 ml of water. The initial yellow color of the solution is discharged by the addition of solid sodium bisulfite in small portions. The mixture is allowed to stand for 1 hour, and the product is collected by suction filtration. Recrystallization from 60–90° petroleum ether affords white needles, mp 78–87° (mixture of *dl* and *meso*), about 24 g, and a second crop of about 3 g. (This product may cause allergic reactions and gloves are recommended.)

2. *Diphenylcyclopropenone:* A 500-ml one-necked flask is fitted with a magnetic stirrer and a pressure-equalizing dropping funnel. The flask is charged with a solution of triethylamine (25 ml) in methylene chloride (65 ml). A solution of the dibromoketone (27 g, 0.073 mole) in 125 ml of methylene chloride is then added dropwise over 30 minutes followed by 30 minutes additional stirring. The mixture is then washed twice with 40-ml portions of 3 N hydrochloric acid (aqueous phase discarded) and the red organic solution is transferred to a 500-ml Erlenmeyer flask. The flask is cooled in an ice bath, and a cold mixture of 12.5 ml of concentrated sulfuric acid and 6.3 ml of water is slowly added with swirling, resulting in the separation of solid diphenylcyclopropenone bisulfate. (If no precipitate appears after 30 minutes, 1-ml portions of concentrated sulfuric acid are added with swirling until solid appears.) The precipitate is collected on a sintered-glass funnel and washed twice with 25-ml portions of methylene chloride. The solid is transferred to a 500-ml Erlenmeyer flask to which is then added 65 ml of methylene chloride, 125 ml of water, and 1.3 g of sodium carbonate in small

portions. The organic layer is separated, the aqueous layer is extracted twice with 40-ml portions of methylene chloride, and the combined organic solutions are dried (anhydrous magnesium sulfate). Evaporation of the solvent (rotary evaporator) affords the crude diphenylcyclopropenone. The crude product is boiled with 75 ml of cyclohexane and the cyclohexane is decanted from the oily residue. This procedure is repeated four more times. On cooling the cyclohexane solution (about 375 ml total), one obtains the product as white crystals, mp 119–120°, about 7.5 g.

IV. Phenylcyclopropane from Cinnamaldehyde

The preparation of cyclopropane derivatives has been greatly facilitated by the development of carbene-type intermediates (see Chapter 13) and their ready reaction with olefins. The preparation of phenylcyclopropane from styrene and the methylene iodide–zinc reagent proceeds in only modest yield, however, and the classical preparation of cyclopropane derivatives by the decomposition of pyrazolines (first employed by Buchner in 1890) is therefore presented in the procedure as a convenient alternative.

PHENYLCYCLOPROPANE (5)

$$C_6H_5-CH=CHCHO + N_2H_4 \cdot H_2O \xrightarrow{\text{EtOH}} \quad C_6H_5-\underset{\underset{H}{|}}{\overset{}{N}}-N \xrightarrow[\text{-N}_2]{\Delta} C_6H_5-\triangleleft$$

Caution: This reaction should be carried out behind a safety screen.

A 500-ml, three-necked, round-bottom flask is equipped with a condenser, a dropping funnel, and a thermometer in the reaction mixture. In the flask is placed a mixture of 85% hydrazine (115 ml, 118 g) and 225 ml of 95% ethanol with a few boiling chips. The solution is brought to reflux (mantle) and cinnamaldehyde (100 g, 0.76 mole) is added dropwise over about 30 minutes followed by an additional 30 minutes of refluxing. A still head is attached to the flask and volatiles (ethanol, water, hydrazine hydrate) are slowly distilled at atmospheric pressure until the pot temperature reaches 200° (about 3 hours). Hereafter, phenylcyclopropane is collected over the range 170–180°. When the pot temperature exceeds 250°, the recovery is complete.* The crude product (55–65 g) is washed twice with 50-ml portions of water and dried (anhydrous potassium carbonate). Distillation under vacuum through a short column affords the product, bp 60°/13 mm, 79–80°/37 mm, n_D^{25} 1.5309, about 40 g (45%).

---

* The cooled pot residue is removed by heating with DMF overnight on a steam bath.

V. Conversion of Alkyl Chlorides to Nitriles in DMSO

Chain extension by means of the reaction of alkyl halides with cyanide is frequently alluded to but rarely employed, mainly because of the long reaction times and poor yields usually encountered. The use of DMSO as a solvent has greatly simplified the procedures and improved the yields of many ionic reactions, and the conversion of alkyl chlorides to nitriles is a good example.

General Procedure and Examples (6)

$$R—Cl + NaCN \xrightarrow{\text{DMSO}} R—CN + NaCl$$

DMSO is dried over calcium hydride and distilled, bp 64°/4 mm, before use. Sodium cyanide is dried at 110° for 12 hours and stored in a tightly stoppered bottle.

1. *Primary Chlorides:* Dry sodium cyanide (30 g, 0.61 mole) is added to 150 ml of dimethyl sulfoxide in a flask fitted with a stirrer, reflux condenser, dropping funnel, and thermometer. The thick slurry is heated on a steam bath to 90° and the steam bath is then removed. The halide (0.5 mole of monochloride or 0.25 mole of dichloride) is slowly added to the stirred mixture, causing the temperature to increase immediately. The rate of addition should be adjusted so that the temperature of the reaction does not go above about 160°. After all the halide is added (about 10 minutes) the mixture is stirred for 10 minutes more, or until the temperature drops below 50°. In the preparation of mononitriles, the reaction mixture is then poured into water, and the product is extracted with chloroform or ether. The extract is washed several times with saturated sodium chloride solution then dried over calcium chloride, and the product is distilled.

With dinitriles a slightly different procedure is necessary due to their water solubility. After the reaction has cooled, 150 ml of chloroform is added to the flask, and this mixture is then poured into saturated salt solution. Enough water is added to dissolve precipitated salt and the chloroform layer is separated. The aqueous layer is extracted once with chloroform. The combined extracts are then washed twice with salt solution, dried, and distilled.

2. *Secondary Chlorides:* With a low-boiling chloride such as 2-chlorobutane, a stirred slurry of 30 g (0.61 mole) of sodium cyanide in 150 ml of dimethyl sulfoxide is heated to 90° with a heating mantle, and 0.5 mole of the chloride is slowly added over a period of 30 minutes. The temperature of the refluxing reaction mixture slowly increases as nitrile is formed. Refluxing continues as the temperature slowly rises to 150° after 3 hours reaction time. The flask is then cooled and the reaction mixture is worked up in the same way as for the primary nitriles. With 2-chlorooctane, the sodium cyanide–dimethyl sulfoxide slurry is heated to 130° and 0.5 mole of the chloride added. The reaction mixture is maintained at 135–145° for 1 hour, then cooled, and the product is isolated as above. Examples are given in Table 16.1.

TABLE 16.1

	Reaction time (minutes)	Product	Yield (%)	bp (°C)/mm	n_D^{25}
1,2-Dichloroethane	20	Succinonitrile	56	114/3.4	—
1,3-Dichloropropane	30	Glutaronitrile	67	101–102/1.5	1.4339
1,4-Dichlorobutane	30	Adiponitrile	88	115/0.7	1.4369
1,5-Dichloropentane	30	Pimelonitrile	75	149/1.0	1.4398
1-Chlorobutane	20	Valeronitrile	93	139/760	1.3949
1-Chloropentane	20	Capronitrile	97	80/48	1.4050
1-Chlorohexane	20	Heptanonitrile	91	96–97/50	1.4125
1-Chlorodecane	20	Undecanonitrile	94	87–88/1.2	1.4314
2-Chlorobutane	180	2-Cyanobutane	69	125–126/760	1.3873
2-Chlorooctane	60	2-Cyanooctane	70	88/12	1.4179

VI. Photolytic Addition of Formamide to Olefins

Free radical additions to mono-olefins are quite common and can frequently be employed to advantage on a synthetic scale. Formamide, for example, on exposure to sunlight or UV radiation adds to olefins in an anti-Markovnikov sense giving 1:1 adducts that are readily isolated and crystallized. Moreover, since alkyl formamides may be conveniently converted to carboxylic acids by conventional means, the reaction represents a general method of chain extension.

ALKYL FORMAMIDES FROM OLEFINS: CATALYSIS BY ULTRAVIOLET LIGHT OR SUNLIGHT (7)

$$R\text{—}CH\text{=}CH_2 + HCONH_2 \xrightarrow[\text{acetone}]{h\nu} RCH_2CH_2CONH_2$$

Formamide is distilled under vacuum before use. t-Butyl alcohol may be dried by distillation from sodium or calcium hydride (3 g/100 ml). All other reagents should be dry.

1. *Irradiation with UV Light:* A photolytic reaction vessel consisting of a high-pressure mercury vapor lamp fitted into a Pyrex tube and immersed in the reaction vessel is employed. The reaction vessel is cooled by running water (to maintain a pot temperature of 30–32°), and nitrogen is passed through the vessel throughout the irradiation (see Appendix 3 for a description of a typical photolytic reaction vessel).

A mixture of olefin (0.005 mole), formamide (40 g), t-butyl alcohol (35 ml), and acetone (5 ml) is irradiated for 45 minutes. A solution of olefin (0.045 mole), t-butyl alcohol (10 ml), and acetone (7 ml) is then added in ten equal portions at 45-minute intervals. The irradiation is continued for an additional 6 hours. At the conclusion of the irradiation, the volatile solvents are removed (rotary evaporator), and the formamide is distilled from the mixture at reduced pressure (approx. 0.2 mm). The residue is dissolved in acetone and filtered to remove traces of oxamide. The acetone is removed, and the residue is treated with water or dissolved in acetone–petroleum ether to induce crystallization. The crude product may be recrystallized from the same solvent.

By this procedure, the following amides (Table 16.2) may be prepared from the indicated starting materials in the approximate yields given.

TABLE 16.2

Olefin	Product	Yield (%)	mp (°C)
1-Heptene	Octanamide	60	105–106
1-Octene	Nonanamide	50	99–100
1-Decene	Undecanamide	70	99–100
Norbornene	*exo*-Norbornane-2-carboxamide	–	181–183

2. *Photolytic Reaction in Sunlight:* A mixture of olefin (0.01 mole), formamide (40 g), *t*-butyl alcohol (20 ml), and acetone (5 ml) is placed in a Pyrex Erlenmeyer flask, which is then flushed with nitrogen, stoppered, and situated in direct sunlight for 1 day. A solution of olefin (0.04 mole), *t*-butyl alcohol (25 ml), and acetone (5 ml) is then added in four equal portions at 1-day intervals, and finally, the flask is left in sunlight for an additional 2 days. Work-up of the solutions as in the above procedure gives the desired amide in comparable yields.

VII. Intermolecular Dehydrohalogenation

Triethylamine has been found to dehydrohalogenate acid chlorides having α-hydrogens to give ketenes according to the reaction. In an interesting variation on this

$$R—CH_2COCl + Et_3N \rightarrow R—CH=C=O + Et_3N·HCl$$

process, cyclohexylcarbonyl chloride is treated with triethylamine to give the product derived from intermolecular dehydrohalogenation.

DISPIRO[5.1.5.1]TETRADECANE-7,14-DIONE (8)

A 500-ml, three-necked, round-bottom flask is equipped with a mechanical stirrer, a condenser (drying tube), a pressure-equalizing dropping funnel, and a nitrogen inlet. The flask is charged with a solution of cyclohexanecarbonyl chloride (30 g, 0.205 mole) in 250 ml of dry benzene, the system is flushed with nitrogen, and a nitrogen atmosphere is maintained thereafter. In the dropping funnel is placed anhydrous triethylamine (35 g, 0.35 mole), which is added to the stirred reaction mixture dropwise (approx. 30 minutes). The mixture is brought to reflux and refluxing is continued (with stirring) 18–20 hours. The cooled solution is then filtered to remove triethylamine hydrochloride, and the filtrate is transferred to a separatory funnel and washed with 200 ml of 5% aqueous hydrochloric acid followed by an equal volume of water. The organic solution is dried (sodium sulfate), the benzene is removed (rotary evaporator), and the residue is recrystallized from petroleum ether–ethanol. The weight of product thus obtained is about 12 g (54%), mp 161–162°.

VIII. Ring Enlargement with Diazomethane

Diazomethane reacts with ketonic compounds according to the reactions to give epoxides and homologated ketones. If the ketone employed is cyclic, only a single

$$R-\underset{\underset{O}{\|}}{C}-R' \xrightarrow{CH_2N_2} RCH_2-\underset{\underset{O}{\|}}{C}-R' + R-\underset{\underset{O}{\|}}{C}-CH_2R' + R-\underset{\triangle_O}{C}-R'$$

homologation product results (plus a small percentage of twice homologated material), and the amount of epoxide formed is quite low (9). In the procedure, cyclohexanone is treated with diazomethane giving a mixture of cycloheptanone, cyclooctanone, and the corresponding epoxide. For the isolation of pure cycloheptanone, advantage is taken of the fact that it is the only product that forms a solid bisulfite addition compound.

CYCLOHEPTANONE FROM CYCLOHEXANONE (10)

This experiment must be carried out in an efficient hood since diazomethane is highly toxic. Also, a safety screen is recommended. (See Chapter 17, Section III for other precautions.)

A 1-liter three-necked flask equipped with a thermometer, a mechanical stirrer and a dropping funnel is charged with 49 g (0.5 mole) of redistilled cyclohexanone, 125 g (0.58 mole) of N-methyl-N-nitroso-p-toluenesulfonamide ("Diazald," Aldrich Chemical Co.), 150 ml of 95% ethanol and 10 ml of water. The nitrosamide is largely undissolved. The height of the stirrer is adjusted so that only the upper part of the solution is stirred and the precipitate moves slightly; the thermometer bulb dips in the liquid. The mixture is cooled to about 0° in an ice–salt bath. While stirring gently, a solution of 15 g of potassium hydroxide in 50 ml of 50% aqueous ethanol is added dropwise very slowly from the dropping funnel. (After 0.5–1 ml of the solution has been added, a vigorous evolution of nitrogen begins and the temperature rises.) The rate of addition is adjusted so that the temperature is maintained at 10–20°. The addition of alkali requires about 2 hours and the nitroso compound is finally consumed. The orange-yellow solution is stirred for a further 30 minutes, and 2 N hydrochloric acid is then added until the solution is acidic to litmus paper (approx. 50 ml).

A solution of 100 g of sodium bisulfite in 200 ml of water is added, and the stirring is continued for 10 hours with exclusion of air. A thick precipitate separates after a few minutes. The bisulfite compound is collected by suction filtration, washed with ether until colorless, and then decomposed in a flask with a lukewarm solution of 125 g of sodium carbonate in 150 ml of water. The ketone layer is separated, and the aqueous layer is extracted four times with 30-ml portions of ether. The combined organic layers are dried over anhydrous magnesium sulfate, the ether is removed at atmospheric pressure, and the residual oil is fractionated under reduced pressure through a short column. The cycloheptanone, bp 64–65°/12 mm, is obtained in about 40% yield.

IX. Conjugate Addition of Grignard Reagents

The addition of Grignard reagents to α,β-unsaturated ketones gives mixtures resulting from 1,2-addition and 1,4-addition. In the presence of cuprous salts, however, the conjugate (1,4) addition is enhanced to the extent that the reaction becomes synthetically useful (*11*). Two examples of this procedure are given.

A. 3,3,5,5-Tetramethylcyclohexanone (*11*)

A 250-ml, three-necked, round-bottom flask is fitted with a mechanical stirrer, a condenser, a nitrogen inlet and outlet, and a pressure-equalizing addition funnel. The flask is charged with 20 ml of anhydrous ether and 3 g (0.12 g-atom) of magnesium turnings. A nitrogen atmosphere is established and maintained in the system. The addition funnel is charged with 19.5 g (0.14 mole) of methyl iodide, and a few drops are added to the stirred flask. After the spontaneous reaction begins, the remainder of the methyl iodide is added at a rate so as to maintain a gentle reflux. After the methyl iodide has been added and the magnesium is completely dissolved, cuprous bromide (0.2 g) is added and the flask is cooled with stirring in an ice-water bath to 5°. A solution of isophorone (11 g, 0.08 mole) in 15 ml of ether is added to the stirred cooled flask over a period of about 20 minutes. The flask is allowed to stir at room temperature for 1 hour, and finally, the mixture is refluxed for 15 minutes. To the cooled reaction mixture is added a mixture of 50 g of ice and 6 g of acetic acid. The ether layer is separated, and the aqueous layer is extracted with an additional 25-ml portion of ether. The combined ether extracts are washed twice with sodium carbonate solution and twice with water. After drying (anhydrous sodium sulfate) and filtration, the ether is removed (rotary evaporator). The residual oil is fractionally distilled yielding a small forerun (approx. 7%) of 1,3,5,5-tetramethyl-1,3-cyclohexadiene, bp 24–25°/7 mm (151–155°/760 mm). The desired product, 3,3,5,5-tetramethylcyclohexanone, is obtained in about 80% yield, bp 59–61°/5.5 mm (196–197°/760 mm).

B. 9-METHYL-cis-DECALONE-2 (12)

By a procedure analogous to that described in the preceding experiment, $\Delta^{1(9)}$-octalone-2 (12 g, 0.08 mole, Chapter 9, Section III) in ether is added to methylmagnesium iodide in the presence of cuprous bromide (0.2 g). After decomposition with ice–acetic acid, extraction with ether, and washing of the ether extract, the ethereal solution is shaken with an equal volume (50–60 ml) of saturated aqueous sodium bisulfite for 3 hours. The mixture is filtered and the filtrate is reserved. The crystals are washed with ether. The filtrate is separated and the aqueous phase is combined with the filtered solid. The combination is acidified (dilute hydrochloric acid) and heated under reflux for 30 minutes. The product thus liberated is extracted into ether, the ether is washed with bicarbonate, then with saturated aqueous sodium chloride solution, and then dried and evaporated. The residual oil is the desired product, bp 250–254°.

X. Dimethyloxosulfonium Methylide in Methylene Insertions

Treatment of trimethyloxosulfonium iodide with sodium hydride generates dimethyl-oxosulfonium methylide according to the reaction. This reagent has the remarkable

$$(CH_3)_3\overset{\oplus}{\underset{O}{S}}I^{\ominus} + NaH \xrightarrow{\text{DMSO}} (CH_3)_2\underset{O}{S}{=}CH_2 + NaI + H_2\uparrow$$

capability of stereospecific reaction with certain α,β-unsaturated carbonyl compounds to form cyclopropane derivatives. For example, it reacts with ethyl *trans*-cinnamate to give ~99% ethyl *trans*-2-phenylcyclopropanecarboxylate in modest yield (*13*). Another reaction of the reagent, methylene insertion, occurs with simple ketones, an example of which is given in the procedure.

METHYLENECYCLOHEXANE OXIDE (*14*)

$$(CH_3)_3\overset{\oplus}{\underset{O}{S}}I^{\ominus} + NaH \xrightarrow{\text{DMSO}} (CH_3)_2\underset{O}{S}{=}CH_2 + NaI + H_2$$

Note: Because hydrogen is generated, the reaction should be carried out in a hood.

1. *Dimethyloxosulfonium Methylide:* A 500-ml, three-necked, round-bottom flask is fitted with a magnetic stirrer, a condenser, a nitrogen inlet and outlet, and a 5-cm length of Gooch (wide diameter rubber) tubing connected to a 125-ml Erlenmeyer flask (see Appendix 3 for an illustration of this setup). In the reaction flask is placed a 50% mineral oil dispersion of sodium hydride (10.5 g, 0.22 mole). The mineral oil is removed by briefly stirring the dispersion with 150 ml of 30–60° petroleum ether, allowing the hydride to settle, and removing the solvent with a pipet. The hydride is then covered with 250 ml of dry DMSO and the 125-ml Erlenmeyer flask charged with 50.6 g (0.23 mole) tri-methyloxosulfonium iodide. The system is flushed with nitrogen and a nitrogen atmosphere is maintained thereafter. Stirring is begun, and the oxosulfonium iodide is added in small portions over about 15 minutes. After the addition, stirring is continued for 30 minutes.

2. *Methylenecyclohexane Oxide:* The Gooch tubing and Erlenmeyer flask are removed and a pressure-equalizing dropping funnel charged with 19.6 g (0.2 mole) of cyclo-hexanone is attached to the reaction flask. The cyclohexanone is added with stirring over about 5 minutes and stirring at room temperature is continued for 15 minutes followed by stirring at 55–60° (water bath) for 30 minutes. The reaction mixture is then poured into 500 ml of water and extracted three times with 100-ml portions of ether. The combined extracts are washed with 100 ml of water followed by 50 ml of saturated sodium chloride solution, and dried (anhydrous sodium sulfate). The ether is removed

by fractional distillation at atmospheric pressure (steam bath), and the residue is transferred to a 50-ml round-bottom flask and distilled under vacuum. Methylene-cyclohexane oxide is collected at 61–62°/39 mm, n_D^{23} 1.4485, about 16 g (71 %).

XI. Acylation of a Cycloalkane: Remote Functionalization

Treatment of Decalin with acetyl chloride and aluminum chloride in ethylene chloride as solvent gives a complex mixture of products as shown (*15*). By variation of the reaction parameters, however, it is possible to maximize the yield of the remarkable reaction product, 10 β-vinyl-*trans*-Decalin 1β,1'-oxide (5). This vinyl ether undoubtedly

arises out of a complex series of hydride transfer reactions, and a possible sequence is shown. In any case, the reaction is an intriguing example of the functionalization of a remote position by an (undoubtedly) intramolecular pathway.

The vinyl ether is quite stable and can be converted readily to a variety of disubstituted trans-Decalins. Its preparation and conversion to a derivative are given in the procedures.

A. 10 β-VINYL-*trans*-DECALIN 1β,1'-OXIDE (*15*)

A 500-ml three-necked flask is fitted with a mechanical stirrer, thermometer, and a dropping funnel, and all openings are protected by drying tubes. Ethylene chloride (120 ml) is placed in the flask and anhydrous aluminum chloride (50 g, 0.38 mole) is added in several portions with stirring. The flask is cooled to maintain the temperature at approx. 25°, and acetyl chloride (47 g, 0.6 mole) is gradually added over about 10 minutes. The stirred solution is now cooled to 5–10° and Decalin (technical grade, 36 g, 0.26 mole) is added slowly over 2 hours. The solution is maintained at 5–10° for 3 hours with continued stirring after the addition is completed. The mixture is now added gradually to a vigorously stirred slurry of crushed ice and water (600 g). The organic (lower) layer is separated, and the aqueous phase is extracted once with 100 ml of ethylene chloride. The combined ethylene chloride solutions are washed several times with cold water, and finally dried over potassium carbonate. The solvent is evaporated (rotary evaporator) and the residue is fractionally distilled. The forerun consists of unchanged Decalin, bp 65–70°/10 mm (185–190°/1 atm). The product is collected at bp 105–111°/8 mm and the yield is about 11 g.

The vinyl ether may be further purified by dissolving it in 15 ml of dry ether and adding a solution of 0.25 g of lithium aluminum hydride in 10 ml of dry ether. The mixture is refluxed for 30 minutes, and excess hydride is destroyed by addition of ethyl acetate (1 ml). Ice-cold dilute (0.5 N) sulfuric acid (25 ml) is gradually added to the cooled mixture, the ethereal layer is rapidly separated, the aqueous layer is extracted once with 10 ml of ether, and the combined ethereal solution is washed once with water and dried over potassium carbonate. Removal of the solvent, followed by distillation of the residue affords about 85% recovery of the pure vinyl ether, bp 102–103°/6 mm, n_D^{25} 1.5045.

B. 10 β-ACETYL-*trans*-1β-DECALOL (*15*)

The vinyl ether prepared above (4.5 g) is dissolved in 30 ml of ether and mixed with 60 ml of 1 N sulfuric acid in a 250-ml flask equipped with a magnetic stirrer and a condenser. The mixture is gently refluxed with stirring for 2.5 hours, and the cooled ether layer is then separated. The aqueous layer is extracted with two 30-ml portions of ether, the combined ether solutions are dried over potassium carbonate, and the ether is removed (rotary evaporator). The residue is recrystallized from petroleum ether, affording the product as prisms, mp 62–64°.

XII. The Modified Hunsdiecker Reaction

The Hunsdiecker reaction is the treatment of the dry silver salt of a carboxylic acid with bromine in carbon tetrachloride. Decarboxylation occurs, and the product isolated is the corresponding organic bromide (16). Since dry silver salts are tedious to prepare, a modification of the reaction discovered by Cristol and Firth (17) is now

$$R—COOAg \xrightarrow{Br/CCl_4} R—Br + CO_2 + AgBr$$

usually employed. The free carboxylic acid is dissolved in carbon tetrachloride, and red mercuric oxide is added. The slow addition of bromine in carbon tetrachloride brings about decarboxylation, and the organic bromide may be isolated in yields comparable to those obtained by the unmodified technique. The procedure given below is typical and may be applied to virtually any carboxylic acid.

1-BROMOHEXANE FROM HEPTANOIC ACID (18)

$$CH_3(CH_2)_5COOH \xrightarrow[Br_2/CCl_4]{HgO} CH_3(CH_2)_5Br + CO_2 + HgBr_2$$

A 500-ml three-necked flask is fitted with a dropping funnel, a condenser, and a magnetic stirrer and arranged in a water bath (or other heating device) for gentle heating (40–50°). In the flask is placed a solution of 13 g (0.1 mole) of heptanoic acid in 150 ml of dry carbon tetrachloride, and 22 g (0.1 mole) of red mercuric oxide. With openings protected by drying tubes, stirring is commenced and the flask is heated to 40–50°. A solution of 16 g (0.1 mole) of bromine in 50 ml of dry carbon tetrachloride is placed in the addition funnel and a few drops are added to the stirred warm solution. An immediate reaction should occur as evidenced by the rapid generation of carbon dioxide and the disappearance of the bromine color. (If the reaction does not occur spontaneously, heating is continued and further addition of bromine is postponed until the reaction starts.) The remainder of the bromine solution is then slowly added (over about 30 minutes) and, finally, the solution is stirred with heating for 1 hour. The cooled reaction mixture is filtered to remove mercuric bromide, and the filtrate is washed with 5% sodium hydroxide followed by water and dried over anhydrous magnesium sulfate. Careful fractional distillation of the solution gives about 6 g (37%) of 1-bromohexane, bp 150–159°.

REFERENCES

1. W. Haaf, *Org. Syn.* **46**, 72 (1966) and references cited therein.
2. P. D. Bartlett, R. E. Pincock, J. H. Rolston, W. G. Schindel, and L. A. Singer, *J. Amer. Chem. Soc.* **87**, 2590 (1965).
3. H. E. Winberg and F. S. Fawcett, *Org. Syn.* **42**, 83 (1962).
4. R. Breslow and J. Posner, *Org. Syn.* **47**, 62 (1967).
5. R. J. Petersen and P. S. Skell, *Org. Syn.* **47**, 98 (1967).
6. R. A. Smiley and C. Arnold, *J. Org. Chem.* **25**, 257 (1960).
7. D. Elad and J. Rokach, *J. Org. Chem.* **29**, 1855 (1964); *J. Chem. Soc.*, p. 800 (1965).
8. N. J. Turro and P. A. Leermakers, *Org. Syn.* **47**, 34 (1967).
9. E. P. Kohler, M. Tishler, H. Potter, and H. T. Thompson, *J. Amer. Chem. Soc.* **61**, 1059 (1939).
10. T. J. de Boer and H. J. Backer, *Org. Syn. Collective Vol.* **4**, 225 (1963).
11. M. S. Kharasch and P. O. Tawney, *J. Amer. Chem. Soc.* **63**, 2308 (1941).
12. A. J. Birch and R. Robinson, *J. Chem. Soc.*, p. 501 (1943).
13. C. Kaiser, B. M. Trost, G. Beson, and J. Weinstock, *J. Org. Chem.* **30**, 3972 (1965).
14. E. J. Corey and M. Chaykovsky, *Org. Syn.* **49**, 78 (1969).
15. G. Baddeley, B. G. Heaton, and J. W. Rasburn, *J. Chem. Soc.*, p. 4713 (1960).
16. C. V. Wilson, *Org. React.* **9**, 332 (1957).
17. S. J. Cristol and W. C. Firth, *J. Org. Chem.* **26**, 280 (1961); J. S. Meek and D. T. Osaga, *Org. Syn.* **43**, 9 (1963).
18. J. A. Davis, J. Herynk, S. Carroll, J. Bunds, and D. Johnson, *J. Org. Chem.* **30**, 415 (1965).

17

Miscellaneous Preparations

I. Derivatives of Adamantane

The preparation of adamantane from readily available starting materials (Chapter 13, Section I) has opened the door to a study of its many substitution products, both from a chemical and a biological point of view (1). Adamantylamine hydrochloride for example, has been found to exhibit antiviral activity. Presented below are several procedures for the preparation of adamantane derivatives.

A. 1-ADAMANTANECARBOXYLIC ACID (2)

Caution: Because carbon monoxide is evolved, the reaction should be carried out in a hood.

A 1-liter, three-necked, round-bottom flask is equipped with a mechanical stirrer, a thermometer immersed in the reaction mixture, a dropping funnel, and a gas vent. In the flask is placed a mixture of 96% sulfuric acid (25.5 ml, 470 g, 4.8 mole), carbon tetrachloride (100 ml), and adamantane (13.6 g, 0.10 mole), and the mixture is cooled to 15–20° with rapid stirring in an ice bath. One milliliter of 98% formic acid is added and the mixture is stirred until the evolution of carbon monoxide is rapid (about 5 minutes). A solution of 29.6 g (38 ml, 0.40 mole) of *t*-butyl alcohol in 55 g (1.2 mole) of 98–100% formic acid is then added dropwise to the stirred mixture over 1–2 hours, the temperature being maintained at 15–20°. After stirring for an additional 30 minutes, the mixture is poured onto 700 g of ice, the layers are separated, and the aqueous (upper) layer is extracted three times with 100-ml portions of carbon tetrachloride. The combined carbon tetrachloride solutions are shaken with 110 ml of 15 *N* ammonium hydroxide, whereupon ammonium 1-adamantanecarboxylate forms as a crystalline solid. This precipitate is collected by filtration through a fritted glass funnel and washed with 20 ml of cold acetone. The salt is transferred to a 500-ml Erlenmeyer flask, mixed with 250 ml of water, and acidified with 25 ml of concentrated hydrochloric acid. Extraction of this mixture with 100 ml of chloroform, drying of the chloroform (anhydrous sodium sulfate), and removal of the chloroform (rotary evaporator) affords a residue of the crude product, mp 173–174°. Recrystallization from methanol–water (3:1) gives about 10 g (56%) of pure 1-adamantanecarboxylic acid, mp 175–177°.

B. 1-BROMOADAMANTANE (3)

Note: The reaction should be carried out in a hood because bromine is employed.

A 100-ml flask is charged with 25 ml of bromine and 10 g of adamantane and heated under reflux for 3 hours. The cooled mixture is dissolved in 100 ml of carbon tetrachloride, and the carbon tetrachloride solution is washed with 100-ml portions of saturated bisulfite solution until the color of bromine is discharged. The solution is then washed twice with water and dried (magnesium sulfate). The solvent is removed (rotary evaporator) and the product is recrystallized from methanol. (For best recovery of the recrystallized material, the methanol solution should be cooled in a Dry Ice cooling bath.) The product has mp 108°.

C. 1-ADAMANTANOL (4)

A 500-ml flask is equipped with a condenser and a magnetic stirrer and charged with 175 ml of water, 18 ml of THF, 10 g of potassium carbonate, 6.5 g of silver nitrate, and 10 g of 1-bromoadamantane. The mixture is heated in a boiling water bath for 1 hour with stirring, cooled, and the crystallized 1-adamantanol is collected by filtration. It may be purified by dissolving it in THF and diluting the solution with water. The product has mp 289–290°.

An alternate procedure has recently been described (5). In a 100-ml flask equipped with a condenser and a magnetic stirrer, 10 g of 1-bromoadamantane is mixed with 15 ml of 0.67 N hydrochloric acid and 13 ml of DMF. The mixture is heated at 105° for

$\frac{1}{2}$ hour with stirring. After cooling, the reaction mixture is diluted with 100 ml of water and filtered by suction. The filtrate is washed several times with water and dried under vacuum to give the product, mp 283–285°, in 95% yield.

D. 2-Adamantanone (5)

A 250-ml glass-stoppered Erlenmeyer flask is fitted with a magnetic stirrer and arranged for heating in an oil bath. The flask is charged with 6.8 g (0.05 mole) of finely divided adamantane and 125 ml of 96% sulfuric acid. A uniform mixture is obtained by rapid magnetic stirring, which is continued throughout the reaction time. The glass stopper is put lightly in place (to allow for the escape of SO_2, when formed) and the flask is heated to 77° for 6 hours. The stirring is interrupted periodically, and the flask is thoroughly shaken to redisperse sublimed adamantane. (The solid should have dissolved completely by about 4–5 hours' reaction time.) The cooled reaction mixture is carefully poured onto ice, and the resulting aqueous mixture is extracted twice with ether. The black polymeric material remaining in the flask is also extracted with ether. The combined ether extracts are washed with saturated sodium chloride solution and dried over sodium sulfate. Evaporation of the solvent gives about 5 g (66%) of crude adamantanone. Purified material, mp 280–282°, may be obtained by steam distillation of the crude material or by direct steam distillation of the aqueous dispersion obtained after quenching. Chromatography through alumina (ether as eluant) provides the pure ketone, mp 285–286°.

II. Percarboxylic Acids

The epoxidation of olefins, as well as other oxidative procedures, require the use of percarboxylic acids. Two of the more easily prepared and more stable compounds are given below.

A. Perbenzoic Acid (6)

$$C_6H_5COOH + H_2O_2 \xrightarrow{CH_3SO_3H} C_6H_5CO_2OH + H_2O$$

Caution: This reaction must be conducted behind a safety shield. By employing a beaker to contain the reactants, destructively high pressures in case of explosion will more readily be vented.

A 500-ml beaker is arranged for the reaction with a mechanical stirrer, a dropping funnel, and an ice-water cooling bath. The beaker is charged with a mixture of 36.6 g (0.30 mole) of benzoic acid and 86.5 g (0.90 mole) of methanesulfonic acid. Hydrogen peroxide (22 g of 70% solution, 0.45 mole) is added dropwise with cooling and rapid stirring over about 30 minutes, during which time the temperature is held at 25–30°. Stirring is continued for 2 hours, and then the solution is brought to 15°. Ice (50 g) is added cautiously followed by 75 ml of ice-cold saturated ammonium sulfate solution, the temperature being held below 25° during the additions. The solution is then extracted three times with 50-ml portions of benzene, and the combined extracts are washed twice with 15-ml portions of cold saturated ammonium sulfate solution and dried over anhydrous sodium sulfate. After filtration, the solution may be used directly for epoxidation (Chapter 1, Section IV) or other oxidation reactions. The conversion to perbenzoic acid is 85–90%, and the resulting solution is about 40% (1.8 M) in perbenzoic acid.

B. Monoperphthalic Acid (7)

A 500-ml, three-necked, round-bottom flask is fitted with a mechanical stirrer, a thermometer, and a wide-stem (powder) funnel. The flask is cooled in an ice–salt bath and charged with 125 ml (approx. 0.5 mole) of 15% sodium hydroxide solution. When the stirred solution reaches −10°, 30% hydrogen peroxide (57.5 g, 52.5 ml, approx. 0.5 mole) previously cooled to −10° is added in one portion. The pot temperature rises and is allowed to return to −10° whereupon 37.5 g (0.25 mole) of phthalic anhydride (pulverized) is added rapidly with vigorous stirring. Immediately upon dissolution of the anhydride, 125 ml (approx. 0.25 mole) of cooled (−10°) 20% sulfuric acid is added in one portion. (The time interval between dissolution of the anhydride and the addition of the cold sulfuric acid should be minimized.) The solution is filtered through Pyrex wool and extracted with ether (one 250-ml portion followed by three 125-ml portions). The combined ethereal extracts are washed three times with 75-ml portions of 40% aqueous ammonium sulfate and dried over 25 g of anhydrous sodium sulfate for 24 hours under refrigeration.

The dried ether solution contains about 30 g (65%) of monoperphthalic acid and is approx. 0.26 to 0.28 M. It may be used directly for oxidation reactions (cf. Chapter 1, Section IV), or stored under refrigeration. Evaporation of the ether under reduced pressure (no heat) affords the crystalline product, mp 110° (dec).

III. Diazomethane

Although diazomethane is both toxic and explosive, its chemical capabilities are so important that it has found widespread use in a variety of organic reactions (see Chapter 7, Section VI and Chapter 16, Section VIII). It may decompose explosively on contact with rough edges (such as boiling chips or ground glass joints), but it can safely be prepared and used when appropriate precautions are observed. The following preparation of diazomethane employs as the starting material N-methyl-N-nitroso-p-toluene-sulfonamide ("Diazald," Aldrich Chemical Co.).

All reactions employing diazomethane must be carried out in a hood behind a safety shield!

DIAZOMETHANE (8)

$$CH_3-\underset{}{\bigcirc}-SO_2-\underset{\underset{NO}{|}}{N}-CH_3 \xrightarrow[\text{aq. KOH}]{\text{EtOH}} CH_3-\underset{}{\bigcirc}-SO_2Et + CH_2N_2$$

The apparatus consists of a 100-ml distilling flask equipped with a dropping funnel and arranged for distillation through an efficient condenser. The condenser is connected to two receiving flasks in series, the second of which contains 20–30 ml of ether. The inlet tube of the second receiver dips below the surface of the ether and both receivers are cooled in ice baths. All connections in the setup are made with bored cork stoppers and all glass tubing is fire polished (Fig. 17.1).

A solution of potassium hydroxide (5 g) in 8 ml of water is placed in the distilling flask and 25 ml of 95% ethanol is added (*no* boiling chips). The flask is heated in a water bath to 65° and a solution of 21.5 g (0.1 mole) of N-methyl-N-nitroso-p-toluene-sulfonamide in 130 ml of ether is added through the dropping funnel over a period of about 25 minutes. The rate of addition should about equal the rate of distillation. When the dropping funnel is empty, another 20 ml of ether is added slowly and the distillation is continued until the distilling ether is colorless. The combined ethereal distillate contains about 3 g (approx. 0.07 mole) of diazomethane and is approximately 0.5 *M*.

For the preparation of an alcohol-free ethereal solution of diazomethane, the generating flask is charged with 35 ml of diethylene glycol monoethyl ether, 10 ml of ether, and a solution of 6 g of potassium hydroxide in 10 ml of water. The flask is heated in a water bath at 70°, and a solution of 21.4 g of Diazald in 140 ml of ether is added over 20 minutes with occasional shaking of the flask. The distillate is collected as above and the yield is the same.

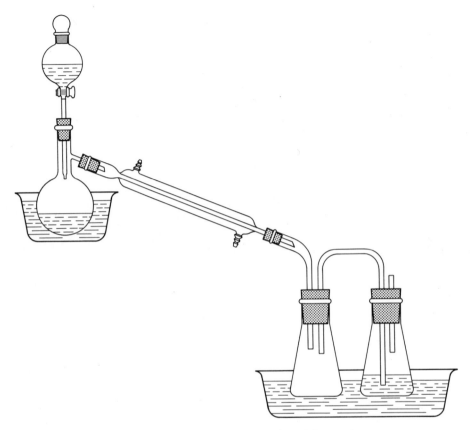

FIG. 17.1. Apparatus for the generation of diazomethane.

IV. Trichloroisocyanuric Acid

Trichloroisocyanuric acid was discovered in 1902 by Chattaway and Wadmore (9). The compound smells like hypochlorous acid and, indeed, has found use commercially as a bleaching agent. It has been used synthetically as an oxidizing agent, and a distinctive use of the reagent is given in Chapter 1, Section VIII wherein an ether is oxidized to an ester. The synthesis given below employing a UV lamp is a modification of that of the discoverers.

TRICHLOROISOCYANURIC ACID (10)

Caution: A free stream of chlorine gas is employed in this reaction, which should therefore be conducted in an efficient hood.

A 500-ml three-necked flask is fitted with a mechanical stirrer, a thermometer, a gas outlet, and a gas inlet tube dipping into the solution. The flask is charged with a solution of cyanuric acid (15 g, 0.116 mole) dissolved in 300 ml of 5% aqueous potassium hydroxide solution. The flask is cooled in an ice–salt bath with stirring to 0° and irradiated with a mercury lamp. A rapid stream of chlorine is passed into the flask (approx. 5 ml/sec), whereupon a heavy white precipitate forms. The addition of gas is continued until the solid material no longer forms (approx. 2 hours). The flask is briefly flushed with air, the product is collected by suction filtration in an ice-cooled funnel, and the residue washed with several small portions of cold water. Since it undergoes slow hydrolysis, the product should be dried in a vacuum oven. The crude product has a variable melting point (195–225°); the yield is about 20 g (approx. 75%).

The crude product (or the commercially available material) may be recrystallized from ethylene chloride giving colorless needles, mp 246–247° (dec) (*11*).

REFERENCES

1. R. C. Fort and P. von R. Schleyer, *Chem. Rev.* **64**, 277 (1964).
2. H. Koch and W. Haaf, *Org. Syn.* **44**, 1 (1964).
3. H. Stetter and C. Wulff, *Chem. Ber.* **93**, 1366 (1960).
4. H. Stetter, M. Schwarz, and A. Hirschhorn, *Chem. Ber.* **92**, 1629 (1959).
5. H. W. Geluk and J. L. M. A. Schlatmann, *Tetrahedron* **24**, 5361 (1968).
6. L. S. Silbert, E. Siegel, and D. Swern, *Org. Syn.* **43**, 93 (1963).
7. H. Bohme, *Org. Syn.* **20**, 70 (1940).
8. T. J. de Boer and H. J. Backer, *Rec. Trav. Chim. Pays-Bas* **73**, 229 (1954).
9. F. D. Chattaway and J. Wadmore, *J. Chem. Soc.* **81**, 200 (1901); cf. E. M. Smolin and L. Rapoport, "s-Triazines and Derivatives," p. 391. Interscience, New York, 1959.
10. T. Ishii, S. Kanai, and T. Ueda, *Yuki Gosei Kagaku Kyokai Shi* **15**, 241 (1957); *Chem. Abstr.* **51** 12107h (1957).
11. R. C. Petterson, U. Grzeskowiak, and L. H. Jules, *J. Org. Chem.* **25**, 1595 (1960).

Appendix 1

Examples of Multistep Syntheses*†

1.

2.

3.

4.

5.

* From commercially available starting materials.
† Numbers over arrows refer to chapter and section in which procedure is discussed.

6.

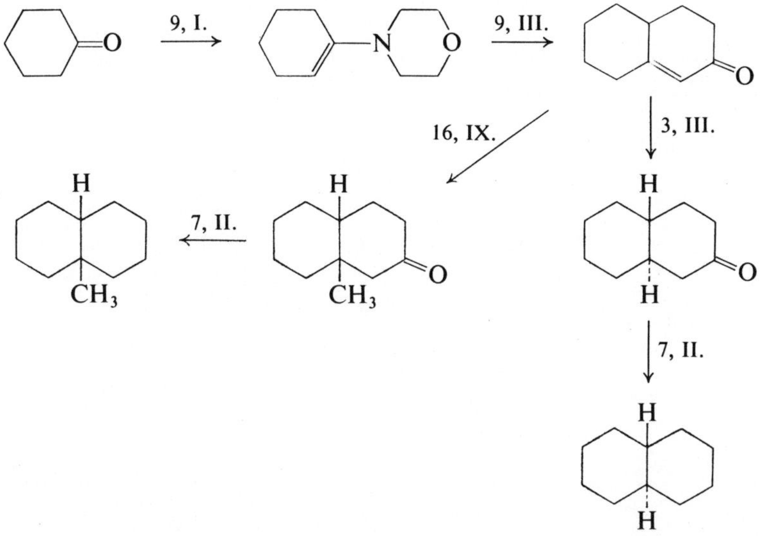

7.

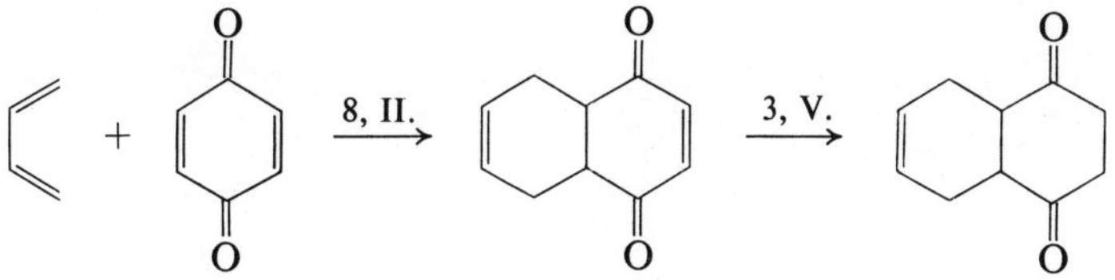

8.

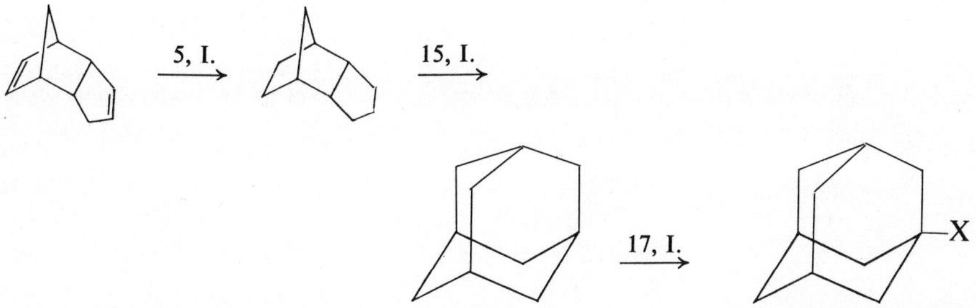

(X = Br, OH, COOH)

9.

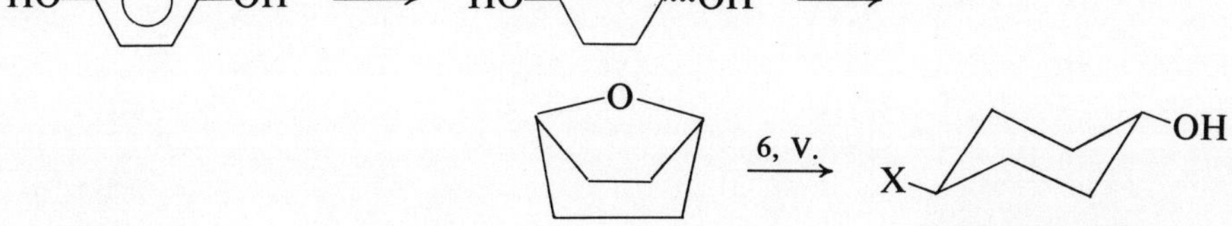

(X = Br, Cl)

10. ϕCHCl + ϕ—CH=CH—CHO $\xrightarrow{11,\ I.}$ [structure] $\xrightarrow{8,\ I.}$ [structure]

11. [structure] + [structure] $\xrightarrow{8,\ I.}$ [structure with COOH, COOH] $\xrightarrow{1,\ X.}$ [cyclohexadiene]

12. [cyclohexanone] $\xrightarrow{9,\ I.}$ [enamine-morpholine] $\xrightarrow{9,\ II.}$

[structure with COCH₃] $\xrightarrow{15,\ V.}$ [cyclopentane-COOH]

13. [cyclohexanone] $\xrightarrow{9,\ V.}$ [enamine-pyrrolidine] $\xrightarrow{9,\ V.}$

[bicyclic structure with N-pyrrolidine and O] $\xrightarrow{9,\ V.}$ [cyclooctene-COOH]

14. [2-methylcyclohexanone] $\xrightarrow{10,\ VI.}$ [decalin structure with CH₃, OH, O] $\xrightarrow{10,\ VI.}$ [octalone with CH₃, O] $\xrightarrow{3,\ III.}$

[decalone with CH₃, H] $\xrightarrow{7,\ II.}$ [decalin with CH₃, H]

[decalone with CH₃, H] $\xrightarrow{6,\ IV.}$ [decalone with CH₃, Br, H]

Appendix 2

Sources of Organic Reagents

The most complete listing of chemicals in commerce is "Chem Sources" published annually by the Directories Publishing Company, Flemington, New Jersey. This volume lists by name all commercially available chemicals with addresses of suppliers.

Of more immediate use for the organic chemist is the remarkable compilation by L. F. Fieser and M. Fieser, "Reagents for Organic Synthesis," Wiley/Interscience. As of this writing, Volume 1 (1967) and Volume 2 (1969) have appeared, and future volumes are anticipated. In their usual thoroughgoing style, the Fiesers have listed hundreds of organic reagents, brief discussions of their sources and applications, and many literature references.

Listed below are the organic compounds required for the synthetic experiments in this volume. The three main suppliers of organic reagents are indicated as follows:

  EK   Eastman Organic Chemicals, Rochester, N.Y.
MCB   Matheson Coleman and Bell, Norwood, Ohio
   A   Aldrich Chemical Company, Milwaukee, Wisconsin

Other suppliers are listed by name.

The reference in parentheses following the name of the compound is to an experimental procedure for the preparation of that compound.

Chapter 1

    I. 4-*t*-Butylcyclohexanol: A
        4-Benzoyloxycyclohexanol (Chapter 7, Section X)
        Chromium trioxide–pyridine complex (Chapter 1, Section I): EK
   II. Camphene: EK, MCB, A
  III. *t*-Butyl perbenzoate: MCB
  IV. Cholesteryl acetate: EK, MCB
   VI. Lead tetraacetate: MCB
       Cyclooctanol: A
  VII. Nitrosyl chloride: Matheson Co.
VIII. Ruthenium dioxide: Alpha Inorganics, Inc.
       Trichloroisocyanuric acid (Chapter 17, Section IV): Chemicals Procurement Laboratories, College Point, N.Y.
  IX. Tetralin: EK, MCB
   X. Lead tetraacetate: MCB
       Dicarboxylic acids (Chapter 8, Sections II and IV)
  XI. Selenium dioxide: MCB

Chapter 2

  I. Diethyl adipate: EK, MCB
     trans-9-Decalincarboxylic acid (Chapter 16, Section I)
     Cyclohexanecarboxylic acid: EK, A
 II. 4-*t*-Butylcyclohexanone (Chapter 1, Section I): A
     1,4-Dioxaspiro[4.5]decane (Chapter 7, Section IX)
III. 4-*t*-Butylcyclohexanone (Chapter 1, Section I): A
     Iridium tetrachloride: Chemicals Procurement Laboratories, College Point, N.Y; D.F.
     Goldsmith Chemical and Metal Corporation, Evanston, Ill.
 IV. Chromium (III) sulfate: MCB
     Diethyl fumarate: EK

Chapter 3

  I. Ethylamine, anhydrous: EK
     Dimethylamine, anhydrous: EK
     Lithium wire: MCB
 II. Ethylenediamine: EK, MCB
     Lithium wire: MCB
III. $\Delta^{1(9)}$-Octalone-2 (Chapter 9, Section III)
     Lithium wire: MCB
     Ammonia, anhydrous: Matheson Co.
     10-Methyl-$\Delta^{1(9)}$-octalone-2 (Chapter 10, Section VI)
 IV. Hexamethylphosphoric triamide (HMPT): EK
     Lithium wire: MCB
     Mesityl oxide: EK, MCB
     Pulegone: A

Chapter 4

  I. Diglyme, bis(2-methoxyethyl) ether: EK, A
     Boron trifluoride etherate: EK, MCB
     Methylenecyclohexane (Chapter 7, Section I; Chapter 11, Sections II and III): A
     4-Methyl-1-pentene: A, MCB
     Norbornene: MCB
     (−)-α-Pinene: A
     cis-2-Butene: Matheson Co.
 II. 2-Methyl-2-butene: EK, MCB
     Boron trifluoride etherate: EK, MCB
     Diglyme: EK, A
     1-Octyne: Chemicals Procurement Laboratories, College Point, N.Y.
     Limonene: EK, MCB
     4-Vinylcyclohexene: MCB, A
III. Octalins (Chapter 3, Sections I and II; Chapter 7, Section III)
     BMB (Chapter 4, Section II)

Chapter 5

  I. Dicyclopentadiene: MCB, EK
     Platinum oxide: MCB
 II. 5% Rhodium on alumina: MCB
III. 5% Rhodium on alumina: MCB

IV. Rhodium chloride: MCB, Alpha Inorganics
Palladium chloride: MCB, Alpha Inorganics
Dowtherm A: Dow Chemical Co.
 V. Rhodium chloride: MCB, Alpha Inorganics
Ergosterol: A, EK, MCB
Geranyl acetate: A
β-Nitrostyrene: A
Methyl oleate: MCB, EK
Naphthoquinone: EK, MCB

Chapter 6

IV. Cholestanone: A
Lithium Carbonate: MCB
 V. 1,4-Cyclohexanediol (Chapter 5, Section II): EK

Chapter 7

 I. *N,N*-Dimethylcyclohexylmethylamine (Chapter 2, Section I)
III. 2-Decalol (decahydro-2-naphthol): A
IV. Boron trifluoride–acetic acid complex: Harshaw Chemical Co., Allied Chemical Co.
 V. 1,5-Cyclooctadiene: A, MCB
Sulfur dichloride: MCB
VI. Boron trifluoride etherate: EK, MCB
VII. Mercuric acetate: MCB
Norbornene: MCB
VIII. Calcium carbide: MCB
Pinacol: EK, MCB
 X. 1,4-Cyclohexanediol (Chapter 5, Section II): EK
XI. Ferrocene: MCB, A
n-Butyllithium: Foote Mineral Co., Alpha Inorganics
Amyl nitrate: K + K Laboratories, Los Angeles
XII. Boron tribromide: MCB
Methyl ethers: A

Chapter 8

 I. 1,4-Diphenyl-1,3-butadiene (Chapter 11, Section I): A
II. 1,3-Butadiene: Matheson Co.
3-Sulfolene (2,5-dihydrothiophene-1,1-dioxide): EK, MCB
IV. Isopropenyl acetate: MCB
Dimedon: EK, MCB
1,3-Cyclohexanedione (Chapter 5, Section II): A
2-Cyclohexen-1-one: A
 V. Dicyclopentadiene: MCB, EK

Chapter 9

 I. Morpholine: MCB, EK
II. 1-Morpholino-1-cyclohexene (Chapter 9, Section I): A
III. Methyl vinyl ketone: MCB, A
1-Morpholino-1-cyclohexene (Chapter 9, Section I): A
IV. β-Propiolactone: EK, MCB
1-Morpholino-1-cyclohexene (Chapter 9, Section I): A
 V. Acrolein: EK, MCB
Pyrrolidine: MCB, EK

Chapter 10

    I.   Sodium hydride, 50% dispersion in mineral oil: Alpha Inorganics
         Ethyl chloroformate: EK, MCB
  III.   Sodium hydride, 50% dispersion in mineral oil: Alpha Inorganics
  VI.   Ethyl 5-bromovalerate: A
         Sodium hydride, 50% dispersion in mineral oil: Alpha Inorganics
         2-Carbethoxycyclooctanone (Chapter 10, Section I)
 VII.   2-Methylcyclohexanone: A, EK
         Methyl vinyl ketone: MCB, A

Chapter 11

   II.   Methyl bromide, ampoule: EK
         n-Butyllithium: Foote Mineral Co., Alpha Inorganics
  III.   Sodium hydride, 50% dispersion in mineral oil: Alpha Inorganics
  IV.   Ethylene oxide: Matheson Co.
   V.   α-Bromo-γ-butyrolactone: A
         Sodium amide: MCB

Chapter 12

    I.   Carbon monoxide: Matheson Co.
         Diglyme: EK, MCB
         Norbornene: MCB
         Boron trifluoride etherate: EK, MCB
   II.   (See I above)
  III.   (See I above)
         Methyl vinyl ketone: MCB, A
         1-Octene: A
  IV.   Potassium *t*-butoxide: MCB

Chapter 13

    I.   Zinc–copper couple (Chapter 13, Section I): Alpha Inorganics
   II.   Potassium *t*-butoxide: MCB
         cis-Cyclooctene: A, MCB
  III.   Sodium trichloroacetate: EK
         Phenylmercuric chloride: A, Alpha Inorganics, MCB
         Phenyl(trichloromethyl)mercury: EK

Chapter 14

    I.   Ammonia, anhydrous: Matheson Co.
         Acetylene: Matheson Co.
   II.   Hexamethylphosphoric Triamide (HMPT): MCB, A, EK
         Acetylene: Matheson Co.
  III.   (See Chapter 10, Section III)
         Triphenylmethane: EK, MCB
         Acetylene: Matheson Co.

Chapter 15

    I.   Tetrahydrodicyclopentadiene (Chapter 5, Section I): A
   II.   Camphor quinone (Chapter 1, Section XI): A
         p-Toluenesulfonylhydrazide: A, EK

III. 1-Ethynylcyclohexanol (Chapter 14, Section I): MCB
Dowex-50 resin: Dow Chemical Co., available from J. T. Baker Chemical Co.
IV. 1,5-Cyclooctadiene: A, MCB
V. 2-Acetylcyclohexanone (Chapter 9, Section II): Chemicals Procurement Laboratories, College Point, N.Y.
VII. Methyllithium, 2 M in ether: Alpha Inorganics

Chapter 16

I. 2-Methylcyclohexanol: EK
Decahydro-2-naphthol: A
II. α-Bromo-p-xylene: A, EK, MCB
Trimethylamine, anhydrous: EK
Phenothiazine: EK, MCB
V. Dibenzyl ketone: EK, MCB
Triethylamine: MCB, EK
VI. Olefins: A
VII. Cyclohexanecarbonyl chloride: EK
Triethylamine: MCB, EK
VIII. N-Methyl-N-nitroso-p-toluenesulfonamide: A
IX. Isophorone: MCB, A
$\Delta^{1(9)}$-Octalone-2 (Chapter 9, Section III)
X. Sodium hydride, 50% dispersion in mineral oil: Alpha Inorganics
Trimethyloxosulfonium iodide (trimethylsulfoxonium iodide): A, EK

Chapter 17

I. Adamantane (Chapter 15, Section I): A
1-Bromoadamantane (Chapter 17, Section I): A
III. N-Methyl-N-nitroso-p-toluenesulfonamide: A
IV. Cyanuric acid: A, EK, MCB

Appendix 3

Introduction to the Techniques of Synthesis

There are some excellent texts and monographs devoted to the techniques of organic synthesis (*1*). Therefore, no attempt will be made in this appendix to be exhaustive. On the other hand, a brief discussion of certain of the more important techniques will allow the student to perform the majority of the included experiments without further exploration.

I. The Reaction

Three operations usually accompany the carrying out of a reaction on a synthetic scale: stirring, addition of a reagent, and temperature control. Most often, a three-necked round-bottom flask allows simultaneous execution of these operations along with certain other controls that may be desirable, such as introduction of an inert atmosphere or maintenance of reflux. In what follows, a short description of a suitable method of carrying out each of the operations is given.

A. STIRRING

Efficient stirring of a reaction mixture is frequently required in carrying out reactions of heterogeneous or viscous mixtures. Moreover, stirring may be necessary during the addition of a reagent to the reaction vessel to prevent high local concentrations or uneven heating or cooling.

Magnetic stirring using a motor-driven magnet in conjunction with a Teflon or Pyrex coated stirring bar is unquestionably the most convenient method of stirring, allowing as it does complete sealing of the reaction vessel. Its use is inadvisable only in the case of highly viscous reaction mixtures or those attended by the presence of large amounts of solid.

When mechanical stirring is required, a glass rod is attached to the drive shaft of an electric motor by a short length of rubber tubing. The rod enters the reaction vessel through a rubber gasket and terminates in a Teflon, Pyrex, or wire stirring paddle. Speed control can be exercised by use of a variable transformer (Fig. A3.1). Although

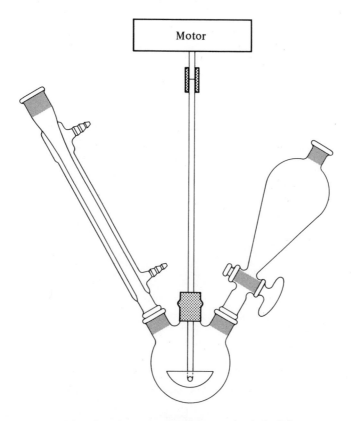

FIG. A3.1. Reaction setup employing mechanical stirring.

this system is not gas-tight, it is usually adequate for those reactions insensitive to air or those conducted under a positive pressure of nitrogen. If total containment of the reaction vessel is necessary, a mercury sealed stirrer (Fig. A3.2a) or a ground glass shaft and bearing stirrer (Fig. A3.2b) must be employed.

B. ADDITION

The addition of liquids or solutions to the reaction vessel can usually be carried out by use of a simple dropping funnel surmounted, if necessary, by a drying tube. For reactions carried out under pressure, vacuum, or an inert atmosphere, a pressure-equalizing dropping funnel is required (Fig. A3.3). Care must be exercised in either case since a single setting of the stopcock will result in a decreased rate of addition as the liquid head diminishes.

If a solid must be added to a closed system over a period of time, an Erlenmeyer flask containing the solid can be attached to the reaction vessel with a short length of

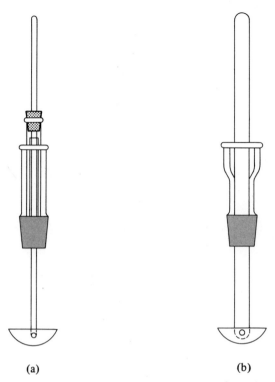

(a)   (b)

FIG. A.3.2. (a) Mercury sealed stirrer (b) Ground glass shaft and bearing stirrer.

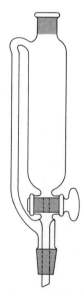

FIG. A3.3. Pressure-equalizing dropping funnel.

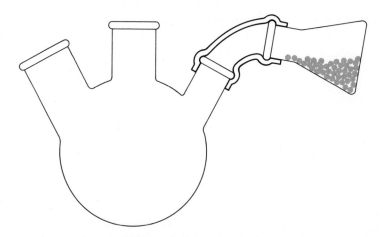

Fig. A3.4. Addition of a solid to a closed vessel.

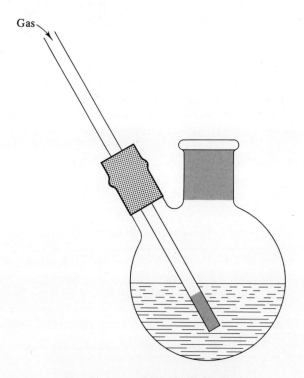

Fig. A3.5. Gas delivery tube in place.

Gooch tubing. Between additions of suitable portions of the solid, the tubing may be sharply bent to protect the solid from the reaction vapors (Fig. A3.4).

Gases may be added above or below the liquid surface by appropriate placement of a fritted-glass gas delivery tube in a rubber gasketed adapter (Fig. A3.5).

C. Temperature Control

The most frequent method of controlling reaction temperature employs a solvent with a suitable boiling point in conjunction with a reflux condenser. Under these circumstances, heating can be done with a steam bath, hot plate, or heating mantle.

When a refluxing solvent cannot be conveniently employed, external control of temperature may be necessary. For reactions carried out below 100°, a water bath heated by a burner or electric hot plate is convenient. Combined hot plate–magnetic stirring units used with a glass or stainless steel water bath allow magnetic stirring through the water bath if desirable. Higher temperatures can be obtained by use of an oil or wax bath with the hot plate. Perhaps the most convenient high temperature heating device, however, is the electric heating mantle. Heating mantle temperatures can be very closely controlled by use of a variable transformer, their internal temperature can be easily monitored by employing the built-in thermocouple, and finally, they are made of nonferrous materials and may be used with magnetic stirring. (On the other hand, they are rather expensive.)

Cooling of the reaction mixture is accomplished by the use of an appropriate cooling bath, the composition of which will vary according to the reaction temperature desired. The baths listed in Table A3.1 may be conveniently contained in a glass, stainless steel, or plastic vessel if magnet stirring is used. (Note, however, that the concentrated HCl–ice mixture should be contained in a glass vessel only.) The Dry Ice baths are best contained in a large Dewar flask, but a simple crystallizing dish is acceptable if magnetic stirring through the vessel is required.

TABLE A3.1
COMMON COOLING BATHS

Coolant	Temperature (°C)
Ice–water	5 to 10
Ice–NaCl (3:1)	−5 to −15
Conc. HCl–ice	−15
CaCl$_2$ (crystal)–ice (3:2)	−40 to −50
Dry Ice–acetone or Dry Ice–isopropyl alcohol	to −80

D. SPECIAL APPARATUS

It is occasionally necessary to carry out special operations in connection with a synthetic procedure. An example frequently encountered is the removal of water from a reaction mixture in order to alter equilibrium concentrations (for example, in the preparations of enamines). For this purpose, a Dean-Stark trap is employed as shown (Fig. A3.6). The reaction is carried out in a solvent that forms an azeotrope with water

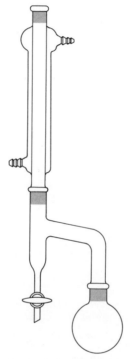

FIG. A3.6. Dean-Stark trap for continuous removal of water.

(such as benzene or toluene). The reaction mixture is refluxed and the distillate condenses and drips into the trap, whereupon it separates into two layers. Water, being more dense than the organic solvent, forms a lower layer. The volume of condensate increases until the organic solvent drips back into the reaction vessel, while the water remains in the trap. Thus, continual azeotropic removal of water is effected with only a relatively small volume of the organic solvent being required.

Another piece of glassware used in synthetic procedures is the cold finger condenser (Fig. A3.7). Used in the case where the reaction is carried out in a low-boiling solvent (for example, liquid ammonia), the condenser is filled with a suitable cooling mixture and fitted with a drying tube.

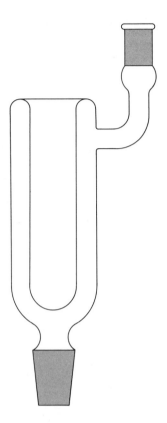

FIG. A3.7. Cold finger condenser.

Reactions under pressure are usually carried out in an autoclave. However, several simple vessels can be used for reactions at moderate pressure. A heavy walled Pyrex test tube or Kjeldahl flask drawn out and sealed with an oxygen torch makes a suitable container for many Diels-Alder reactions. The tube can be heated in an oil or water bath, but care *must* be exercised to protect against explosions. At the conclusion of the reaction, the tube is cooled to room temperature, the neck is scratched with a file or carborundum chip, and a hot Pyrex rod is touched to the scratch. A large crack in the neck should result, and the sealed top can be easily knocked off.

A heavy-walled hydrogenation bottle will tolerate pressures up to 5 atmospheres, although an efficient seal is difficult to devise. A wired-on rubber stopper will serve but will leak well below the limiting pressure tolerance of the bottle.

A heavy-walled Pyrex tube with a suitable lip can be sealed with a metal bottle cap and capper available at many hardware stores (Fig. A3.8). This vessel is suitable for reactions at moderate pressure only (for example, benzene heated to 100°).

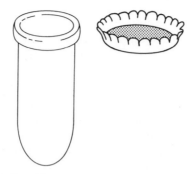

Fig. A3.8. Vessel for reactions under moderate pressure.

E. Hydrogenation

The most convenient setup for hydrogenation on an intermediate synthetic scale is the Parr low-pressure shaker-type apparatus (Fig. A3.9), in which variable pressures of from 1 to 5 atmospheres (60 psi) may be safely employed. The compound to be hydrogenated (approx. 100 g) is dissolved or suspended in 200 ml of a suitable solvent in a heavy-walled 500-ml bottle. The bottle is placed in the apparatus and clamped in place inside the protective mesh. The flask is briefly evacuated, then filled with hydrogen to the desired pressure. The reaction is initiated by starting the shaker, and the course of the reaction is easily followed by observing the drop in pressure in the bottle and

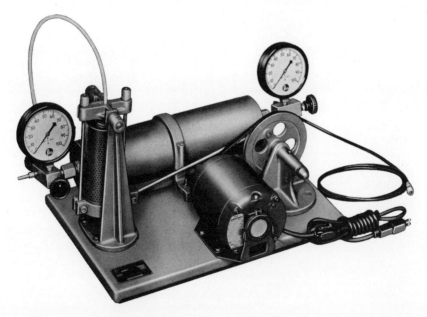

Fig. A3.9. Low-pressure hydrogenation apparatus (Parr Instrument Company).

reservoir tank. The relationship between pressure drop and molar uptake is best established in advance by calibrating the system with a weighed amount of a known olefin (such as cyclohexene) brought up to the recommended total volume (300 ml) with solvent.

At the conclusion of the reaction, the excess hydrogen is vented, the bottle is removed, and the bulk of the catalyst is recovered by filtration. In practice, the catalyst is so finely divided that a second filtration through celite is often necessary to give a pure filtrate. The solvent is then removed and the residue is worked up by standard procedures.

For a discussion of other modes of conducting hydrogenation, the student is referred to the standard reference works on organic methods (*1*).

F. Photolytic Reactions

Reactions catalyzed by light of wavelength greater than 320 mμ may be carried out in Pyrex vessels toward which are directed ordinary sunlamps or sunlight. For reactions requiring light of higher energy, a mercury vapor lamp housed in a quartz or Vycor

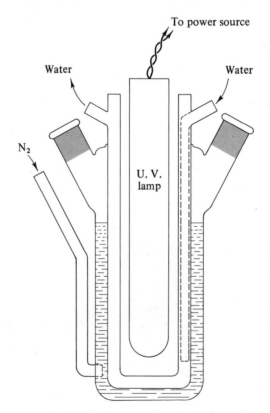

Fig. A3.10. Apparatus for photolytic reactions.

immersion jacket and placed in a suitably constructed reaction vessel is recommended. Since such a lamp generates a considerable amount of heat, a circulating water jacket situated between the lamp and the reaction vessel is desirable. Finally, provision should be made for conducting the reaction under an inert atmosphere (nitrogen or argon) since oxygen interferes with many photolytically initiated free-radical processes. A suitable reaction setup is shown in Fig. A3.10.

II. The Workup

A. EXTRACTION

At the conclusion of the reaction, the resulting complex mixture of reagents and products can frequently be simplified by simple extraction into a suitable solvent, followed by appropriate washings.

If the desired product is fairly water soluble, simple extraction into organic solvents may not be an efficient means of recovery. In that case, continuous extraction of the aqueous solution with an organic solvent may be necessary to effect the recovery. Either of two types of apparatus are normally employed, and the correct design depends on the density of the organic solvent. For solvents less dense than water, the apparatus should be set up as in Fig. A3.11a. The barrel of the extractor is charged with the

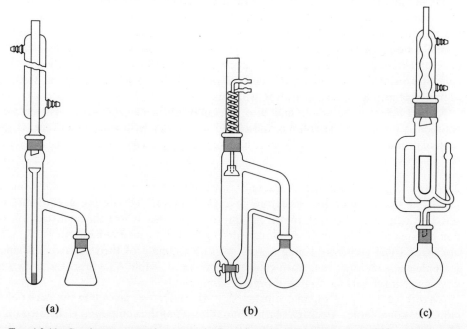

(a)	(b)	(c)

FIG. A3.11. Continuous extraction setups (a) for solvents less dense than water (b) for solvents more dense than water (c) Soxhlet extractor.

aqueous mixture and pure solvent is placed in the flask. The refluxing solvent condenses, drips into the dispersion tube, and bubbles up through the aqueous mixture. The organic solvent accumulates above the aqueous phase until it drips back into the boiling flask. Thus pure solvent is always passing through the aqueous phase, and several hours (usually 18–20) suffices to extract the desired organic compound completely into the boiling flask, even with an unfavorable distribution coefficient.

The second apparatus, shown in Fig. A3.11b, is used with organic solvents more dense than water. The operating principles are identical, except that the return of the solvent occurs from beneath the aqueous mixture.

Finally, the extraction of solid or semisolid masses into solvents can be carried out by use of a Soxhlet extractor (Fig. A3.11b). The sample is placed in a porous cup in the extractor. The boiling solvent condenses into the cup and accumulates until a siphon column is established in the adjacent tube. Then the saturated solvent returns to the boiling flask and fresh solvent distills again, repeating the process.

B. ADSORPTION OF IMPURITIES

1. *Decolorization:* Small amounts of colored impurities are removed from the solution of the organic material by treatment with activated charcoal (Norit, Darco, Nuchar, etc.). The solution is brought to the boiling point and charcoal (1–2 % of the weight of the organic solute) is added. The mixture is filtered hot by gravity into a vessel containing a small amount of the solvent which is kept boiling. The boiling solvent heats the filter cone and inhibits premature crystallization. After the filtration is complete, excess solvent is distilled away and crystallization is induced. Water and most organic solvents are suitable media for decolorization with charcoal.

A short column of alumina may also be employed for decolorization. The colored solution is placed on the column and eluted with a dry hydrocarbon solvent. If the desired product is not highly polar in nature, recovery by the technique may be excellent.

2. *Drying Agents:* Removal of traces of water from organic liquids or solutions is usually desirable prior to distillation or crystallization. The drying agent must be selected with care to ensure that a minimum of product is lost through reaction or complexing, while allowing for adequate drying for the purpose at hand. In Table A3.2 are listed some common drying agents with an estimate of their relative efficiency (completeness of water removal), capacity, and speed. Table A3.3 gives the preferred drying agent for various classes of organic compounds.

Azeotropic drying of organic substances is also effective, providing the material is relatively nonvolatile. A benzene or toluene solution of the compound is distilled (in a Dean-Stark trap, if available) until the distillate is free of water droplets. The remaining solution is essentially dry.

TABLE A3.2
COMMON DRYING AGENTS

Drying agent	Efficiency	Capacity	Speed	Comments
Na	High	Low	Slow	Best for hydrocarbons[a]
P_2O_5	High	Low	Fast	Unsuitable for acid-sensitive compounds; preliminary drying recommended
CaO, BaO	High	Fair	Slow	Best for alcohols, amines, DMF[a]
CaH	High	Fair	Slow	Best for alcohols, DMSO[a]
$CaSO_4$	High	Low	Fast	Preliminary drying recommended
NaOH, KOH	Fair	Low	Fast	Good for amines
$CaCl_2$	Fair	High	Fast	Caution: Forms complexes with alcohols, amines, and many carbonyl compounds
$MgSO_4$	Fair	High	Fast	Good general drying agent
K_2CO_3	Fair	Fair	Slow	Unsuitable for acidic compounds; good for alcohols, amines
Na_2SO_4	Poor	High	Slow	Good preliminary drying agent.

[a] Dried solution may be distilled from drying agent.

TABLE A3.3
PREFERRED DRYING AGENTS FOR COMPOUND CLASSES

Compound class	Recommended drying agent
Hydrocarbons and ethers	$CaCl_2$, $CaSO_4$, Na, P_2O_5, CaH
Alcohols	K_2CO_3, $MgSO_4$, $CaSO_4$, CaO, BaO, CaH
Halides	$CaCl_2$, Na_2SO_4, $MgSO_4$, $CaSO_4$, P_2O_5
Amines	KOH, NaOH, CaO, BaO
Aldehydes	Na_2SO_4, $MgSO_4$, $CaSO_4$
Ketones	Na_2SO_4, $MgSO_4$, $CaSO_4$, K_2CO_3
Acids	Na_2SO_4, $MgSO_4$, $CaSO_4$

C. REMOVAL OF SOLVENTS

The removal of a large volume of solvent prior to distillation or crystallization is frequently necessary. If the solvent and product have similar boiling points (within 75°), careful fractional distillation of the entire solution is required. In other cases, the solvent may be removed by boiling on a steam bath, by rapid distillation at atmospheric pressure, or on a rotary evaporator. The rotary evaporator, used under aspirator pressure in conjunction with a heated water bath, allows rapid removal of most solvents boiling up to about 120°/1 atm (Fig. A3.12a). The presence of a thin film of liquid on the

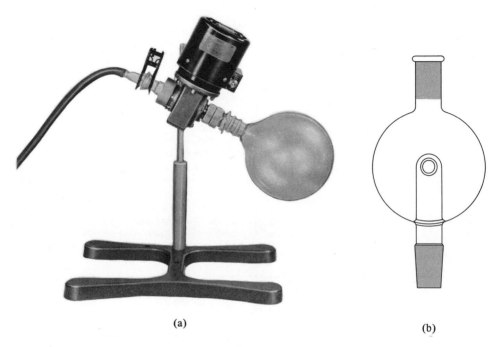

(a)   (b)

FIG. A3.12. (a) Rotary evaporator (Buchler Instruments) (b) Trap used with rotary evaporator.

walls of the rotating flask provides a large surface area for rapid evaporation, while the rotation action mixes the solution and inhibits actual boiling. Figure A3.12b illustrates a trap which may be used with the rotary evaporator to prevent loss of the solution in case of bumping. The trap may also be cooled, if desired, for recovery of the solvent.

III. Purification of the Product

A. DISTILLATION (2)

Setups for simple and fractional distillation at atmospheric pressure are shown (Fig. A3.13). A 30-cm Vigreux column (Fig. A3.13b) is convenient if the components boil at least 50° apart at atmospheric pressure. For better separation, a column packed with glass helices is suitable. All columns employed in fractional distillation should be wrapped or jacketed to minimize heat loss.

Heat sources for distillation must be closely controlled to prevent overheating or too rapid distillation. The best heat sources are electrically heated liquid baths. Mineral oil or wax is a satisfactory medium for heat exchange up to about 240°. The medium may be

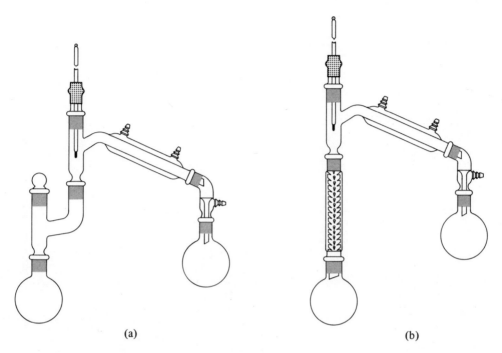

(a)   (b)

FIG. A3.13. Setups for atmospheric pressure distillation (a) for simple distillation (b) Vigreux column for fractional distillation.

contained in a stainless steel beaker or sponge dish and heated by an electric hot plate or immersion coil. The bath temperature (20–80° above the boiling point) is easily monitored by an immersed thermometer.

Distillation at reduced pressure is advisable with the majority of organic compounds boiling above 150° at 1 atmosphere. Aspirator pressure (20–30 mm depending on water temperature and system leaks) is sufficient for many reduced pressure distillations. A liquid boiling at 200°/1 atm, for example, will have a boiling point of approximately 100° at 30 mm. (Estimates of observed boiling points at reduced pressure can be made by use of the pressure-temperature alignment chart shown in Fig. A3.14). The aspirator pump is simple and is not affected by organic or acid vapors. The pressure in such a system is best monitored by a manometer.

A vacuum system employing an oil pump is shown schematically in Fig. A3.15. Protection of the pump requires that the system be well trapped between the pump and the distillation setup. The pressure can be regulated by introducing an air leak through a needle valve (a bunsen burner needle valve is satisfactory). The pressure is monitored by use of a tipping McLeod gauge (Fig. A3.16) which gives intermittant (as opposed to continuous) reading of pressure down to about 0.05 mm, of sufficient precision for the purpose.

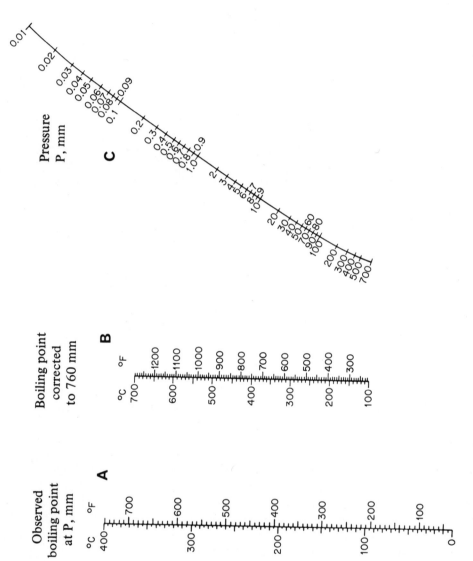

Fig. A3.14. Pressure–temperature alignment chart (reprinted by permission from MC/B Manufacturing Chemists, Norwood, Ohio).

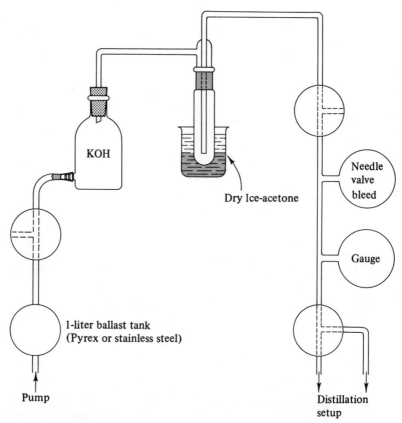

KOH

Dry Ice-acetone

Needle
valve
bleed

Gauge

1-liter ballast tank
(Pyrex or stainless steel)

Pump

Distillation
setup

Fig. A3.15. Schematic diagram of a vacuum system for distillation.

The prevention of bumping in reduced pressure distillations requires special pre-cautions. Boiling chips rarely function well over the course of a long distillation under vacuum, and one of several alternative techniques should be employed. The following methods are listed in decreasing order of effectiveness: (1) Introduction of a fine stream of air or nitrogen through a capillary bleed tube (Fig. A3.17); (2) the use of 12–15 microporous boiling chips (Todd Scientific Co.); (3) covering the boiling liquid with a mesh of Pyrex wool; (4) the use of boiling sticks.

B. CRYSTALLIZATION AND RECRYSTALLIZATION

Several techniques are usually employed to induce crystallization from saturated solutions of organic solids. The introduction of seed crystals will invariably work, although with new compounds such crystals are not available. Seeding with crystals of a

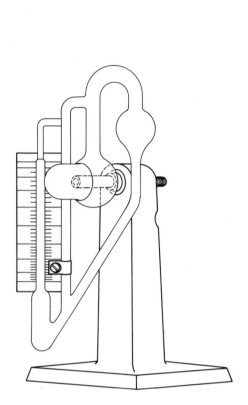

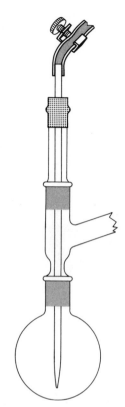

FIG. A3.16. Tipping McLeod guage.

FIG. A3.17. Capillary bleed tube for reduced pressure distillation.

compound with related molecular or crystal structure is frequently successful. Alternatively, cooling the solution and scratching the interior of the vessel with a glass rod is successful in a surprising number of cases.

The technique of trituration is frequently useful. The organic product is stripped of solvent and the oily residue is placed in a mortar and covered with a layer of a solvent in which it is only slightly soluble. The mass is ground with a pestle mixing in the solvent as thoroughly as possible. In favorable cases, the solvent removes traces of impurities that may be inhibiting crystallization, and grinding action induces crystallization.

Successful recrystallization of an impure solid is usually a function of solvent selection. The ideal solvent, of course, dissolves a large amount of the compound at the boiling point but very little at a lower temperature. Such a solvent or solvent mixture must exist (one feels) for the compound at hand, but its identification may necessitate a laborious trial and error search. Solvent polarity and boiling point are probably the most important factors in selection. Benzhydrol, for example, is only slightly soluble in 30–60° petroleum ether at the boiling point but readily dissolves in 60–90° petroleum ether at the boiling point.

Until one develops a "feel" for recrystallization, the best procedure for known compounds is to duplicate a selection in the literature. For new compounds, a literature citation of a solvent for an analogous structure is often a good beginning point. To assist in the search, Table A3.4 lists several of the common recrystallizing solvents with useful data. The dielectric constant can be taken to be a rough measure of solvent polarity.

TABLE A3.4
RECRYSTALLIZING SOLVENTS

Solvent	B.P. (°C)	Dielectric constant	Water solubility (g/100 g)
Acetic acid	188	6.2	Misc.
Acetone	56.5	21	Misc.
Acetonitrile	82	38	Misc.
Benzene	80	2.3	0.07
t-Butyl alcohol	82	17	Misc.
Carbon tetrachloride	77	2.2	0.08
Chloroform	61	4.8	1.0
Cyclohexane	81	2.0	Sl. sol.
DMF	154	38	Misc.
Dioxane	101	2.2	Misc.
DMSO	189	45	Misc.
Ethanol	78	25	Misc.
Ethyl acetate	77	6.0	9
Ethyl ether	35	4.3	7.5
Ethylene chloride	83	10	0.83
Heptane	98	2.0	Insol.
Hexane	69	1.9	Insol.
Isopropyl alcohol	82	18	Misc.
Methanol	65	34	Misc.
Methylene chloride	40	9.1	2.0
Nitromethane	101	38	10
Pentane	36	2.0	0.03
Pyridine	115	12	Misc.
Water	100	80	—

C. DRYING OF SOLIDS

A solid insensitive to air is easily dried by spreading the material over a large piece of filter paper and allowing moisture or solvent to evaporate. However, many organic solids are sensitive to air or moisture and must be dried under reduced pressure in a vacuum desiccator or vacuum oven. Moreover, complete drying of a sample to be analyzed by combustion analysis necessitates vacuum drying. For vacuum drying of small samples, an Abderhalden (drying pistol) is a convenient arrangement (Fig. A3.18).

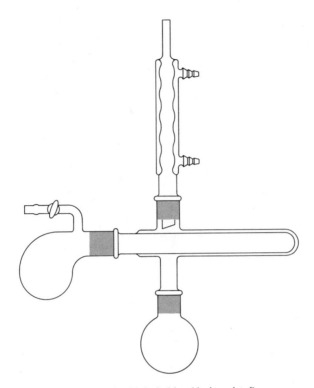

FIG. A3.18. Abderhalden (drying pistol).

The sample is placed in the barrel of the pistol and a drying agent (usually P_2O_5) placed in the "handle." The evacuated system is heated by refluxing a liquid of the desired boiling point over the sample.

D. SUBLIMATION

When a solid compound possesses a relatively high vapor pressure below its melting point, it may be possible to purify it by sublimation. Selenium dioxide, for example, is easily purified prior to use by sublimation at atmospheric pressure (Chapter 1, Section XI). More commonly, the method of choice is sublimation at reduced pressure, which allows more ready evaporation of solids with limited volatility. A convenient vacuum sublimation apparatus is shown in Fig. A3.19. The impure sample is placed in the lower cup, which is attached to the condenser by an O-ring seal and spring. Water is run through the condenser and the system is evacuated. The cup is heated gradually with an oil bath, and sublimation follows. The sublimate is recovered by scraping it off the walls of the condenser with a spatula.

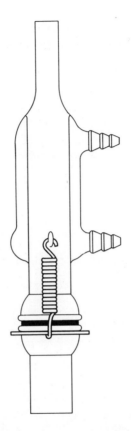

FIG. A3.19. Apparatus for vacuum sublimation.

E. CHROMATOGRAPHY

1. *Column Chromatography:* Column Chromatography is a useful separation technique for mixtures resulting from intermediate to small scale synthetic processes. For example, nitroferrocene is conveniently isolated from a mixture of the product, ferrocene, and 1,1′-dinitroferrocene by chromatography on Activity I basic alumina at about the 100-g scale (Chapter 7, Section XI).

The column (20–30 cm by 1–2 cm diameter or larger in the proportion 10:1) is prepared by filling it with a dry solvent of low polarity (e.g., pentane), pushing a plug of cotton to the bottom, covering the cotton with a layer of sand, and dusting in the adsorbant. About 25 g of adsorbant per gram of mixture is a good approximation for a first trial. The adsorbant is covered with a layer of sand, excess solvent is drained, and the sample, dissolved in a minimum amount of a suitable solvent, is introduced with a dropper.

Alumina is the most frequently employed adsorbant. Its activity (i.e., the extent to which it adsorbs polar compounds) is largely a function of the amount of water present. Alumina of Activity I is prepared by heating the material in an oven to 200–230° and allowing it to cool in a desiccator. Addition of water to the extent of 3%, 6%, 10%, or 15% to the dry material gives alumina of Activity II, III, IV, and V, respectively.

The column is eluted with dry solvents of gradually increasing eluting power. The order of eluting power of the common dry solvents is shown in Table A3.5. The compounds are eluted from the column in order of their increasing polarity. The usual order of elution of organic compounds is shown in Table A3.6. The progress of the

TABLE A3.5
ORDER OF ELUTING POWER OF COMMON DRY SOLVENTS

1. 30–60° Petroleum ether	8. Chloroform
2. 60–90° Petroleum ether	9. Ethyl acetate
3. Carbon tetrachloride	10. Ethylene chloride
4. Cyclohexane	11. Ethanol
5. Benzene	12. Methanol
6. Ether	13. Water
7. Acetone	14. Acetic acid

TABLE A3.6
ORDER OF ELUTION OF ADSORBED COMPOUNDS

1. Aliphatic hydrocarbons	7. Ketones
2. Alkyl halides	8. Aldehydes
3. Olefins	9. Thiols
4. Aromatic hydrocarbons	10. Amines
5. Ethers	11. Alcohols
6. Esters	12. Carboxylic acids

elution is followed by collecting small samples of the eluant and evaporating the solvent. The melting points and spectra of the residual materials serve as a guide to the development of the column. When one compound has been completely eluted, changing to a solvent or a solvent mixture of higher eluting power will hasten the recovery of subsequent fractions.

2. *Thin-Layer Chromatography (TLC):* The function of TLC in organic synthesis is primarily one of allowing the experimenter to follow the progress of the reaction without actually interrupting the reaction. Since successful TLC can be carried out on a minute scale, only a very small fraction of the reaction mixture need be withdrawn and subjected to analysis. The following example of the TLC analysis of the chromic acid oxidation of borneol, described by Davis (*3*), is a useful model.

(a) *Preparation of the plates* (*4*): Microscope slides are washed thoroughly with soap, rinsed with distilled water followed by methanol, and allowed to dry on edge. A suspension of 35 g of Silica Gel G in 100 ml of chloroform is placed in a wide-mouth bottle. Two slides, held face to face with forceps, are immersed in the suspension which is briefly stirred. The slides are withdrawn evenly, resulting in a smooth deposit of the adsorbant. The rate of withdrawal of the slides controls the thickness of the silica gel layer. The slides are separated and allowed to dry. Each slide is then held in a slow stream of steam for 5 seconds to allow the binder to set. Prior to use, the slides are activated by heating for 45 minutes in an oven at 125° or by placing them on a hot plate over a wire gauze for the same length of time.

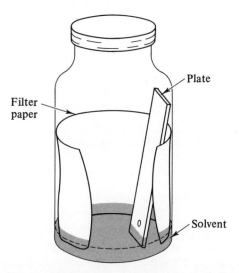

FIG. A3.20. Apparatus for the development of TLC plates.

(b) *Chromic acid oxidation of borneol:* The following solutions are prepared: 2% borneol in ether; 10% chromic anhydride and 5% sulfuric acid in water; 2% camphor in ether.

One milliliter each of the borneol solution and the oxidizing solution are mixed in a test tube and briefly shaken. A TLC slide is spotted with the borneol solution, the camphor solution, and the ether layer of the reaction mixture. Spotting is done by means of a capillary melting point tube used as a dropper and filled with a 5 mm sample. The slide is developed in a wide-mouth jar containing a filter paper liner and a few milliliters of chloroform (Fig. A3.20). After development (the solvent front rises to within 1 cm of the top), the slide is removed, the solvent is allowed to evaporate, and the slide is placed in a covered wide-mouth jar containing a few crystals of iodine. The spots readily become visible and the progress of the reaction can easily be followed. With periodic shaking, the oxidation is complete in about 30 minutes.

A variety of reaction mixtures can be analyzed by this simple technique, although a suitable solvent or solvent mixture for the development of the slide must be determined for the particular compounds involved.

3. *Gas–Liquid Phase Chromatography (glpc):* glpc is certainly a technique of high utility to the synthetic chemist, both for analysis of reaction mixtures and for their separation on a synthetic scale. However, a detailed treatment of the techniques of glpc would be beyond the intention of the present book, since, by and large, such matters as sampling techniques, flow rates, column temperature and packing, as well as other variables, can usually be determined only in connection with the problem at hand. Instead, the student is advised to consult the instruction manuals of individual commercial instruments for operating details. An excellent discussion of the practical aspects of glpc by Ettre and Zlatkis is also available (5). Finally, a useful summary of column packing materials with many references is published periodically by Analabs, Inc. (6).

REFERENCES

1. The two most comprehensive works are A. Weissberger, ed., "Technique of Organic Chemistry." Interscience, New York, 1945; and E. Müller, ed., "Methoden der organische Chemie (Houben-Weyl)," 4th ed. Georg Thieme, Stuttgart. An excellent brief summary of many important techniques is to be found in K. B. Wiberg, "Laboratory Technique in Organic Chemistry." McGraw-Hill, New York, 1960.
2. A thorough discussion of the methods of conducting distillation will be found in A. Weissberger, ed., "Technique of Organic Chemistry," Vol. IV: Distillation. Interscience, New York, 1951. *Cf.* Wiberg, *op. cit.,* Chap. 1.
3. M. Davis, *J. Chem. Educ.* **45**, 192 (1968).
4. J. J. Peiffer, *Mikrochim. Acta,* p. 529 (1962); J. F. Janssen, *J. Chem. Educ.* **46**, 117 (1969).
5. L. S. Ettre and A. Zlatkis, eds., "The Practice of Gas Chromatography." Interscience, New York, 1967.
6. T. R. Lynn, C. L. Hoffman, and M. M. Austin, "Guide to Stationary Phases for Gas Chromatography." Analabs, Inc., Hamden, Connecticut.

Subject Index

A

Abderhalden, 183–184

Acetaldehyde, reaction with sodium acetylide, 124

Acetone, carbethoxylation of, 90

Acetophenone, carboxylation of, 99

1-Acetoxybicycloalkane derivatives, 75–77

endo-2-Acetoxybicyclo[2.2.2]oct-2-ene-5,6-dicarboxylic anhydride, 77

1-Acetoxybicyclo[2.2.2]oct-2-en-5-one, 15, 158

endo-1-Acetoxybicyclo[2.2.2]oct-3-one-5,6-dicarboxylic acid, 15, 77, 158

1-Acetyoxy-8,8-dimethylbicyclo[2.2.2]-oct-2-en-5-one, 15

endo-1-Acetoxy-8,8-dimethylbicyclo[2.2.2]oct-3-one-5,6-dicarboxylic acid, 15, 76

Acetyl chloride, acylation with, 147–148

2-Acetylcyclohexanone, 160, 165
  oxidative rearrangement of, 130–131

1-Acetylcyclohexene, 129–130

10 β-Acetyl-*trans*-1 β-decalol, 148–149

Acetylene, 164
  conversion to sodium acetylide, 121–124
  purification of, 122

Acetylides, *see* Sodium acetylide

Acid chlorides
  dehydrohalogenation of, 142–143
  reaction with enamines, 80

Acrolein, 163
  reaction with enamines, 84–86
  with trialkylboranes, 114

Acrylonitrile, in Diels–Alder reaction, 75

Activated charcoal, 176

Acylation
  of cycloalkanes, 147–148
  of enamines, 81–82

Adamantane, 126–127, 159, 165
  derivatives of, 151–153

1-Adamantanecarboxylic acid, 151–152, 159

1-Adamantanol, 152–153, 159

2-Adamantanone, 153

1,4-Addition, *see* Conjugate addition

Addition funnels, 167–168

Addition of reagents, 167–170

Adiponitrile, 141

Adsorption of impurities, 176–177

Alcohols
  conversion to halides, 45, 46
  dehydration of, 56
  tertiary, esterification of, 62–63

Aldehydes
  by oxidation of primary alcohols, 3, 5
  reaction with acetylides, 124

Alkylation of β-ketoesters, 99–101

Alkyl boranes, 31, 111–115

Alkyl bromides, by triphenylphosphine dibromide, 46–47

Alkyl chlorides
  conversion to nitriles, 140–141
  by triphenylphosphine–carbon tetrachloride, 45–46

Alkyl formamides, by photolytic addition to olefins, 141–142

Alkyl halides
  reaction with acetylides, 121–124
    with enamines, 80
  by triphenylphosphine–carbon tetrabromide, 45
  by triphenylphosphine–carbon tetrachloride, 45–46
  by triphenylphosphine dihalide, 46–47

Alkylidenetriphenylphosphorane, 104–110

β-Alkylpropionaldehydes, from trialkylboranes, 114

Alkynes, by ethynylation, 121–124

Allenes, from 1,1-dihalocyclopropane, 132–133

Allylic bromination, 48–49
  by *N*-bromosuccinimide, 48–49
  of cyclohexene, 48–49
  of 2-heptene, 49

Allylic oxidation, 7–8

Alumina
  activity of, 186
  in decolorization, 176
  in dehydration of 1,4-cyclohexanediol, 51–52

Methyl ethers
  demethylation of, 66–67
  by diazomethane, 59–60
Methyl iodide
  in alkylation of β-ketoesters, 100–101
  in methiodide formation, 85, 86
Methyl isobutyl ketone, 29
Methyl ketones
  by reduction of β-ketosulfoxides, 95
  from trialkylboranes, 114
Methyl linoleate, deuteration of, 44
Methyllithium, 165
  in formation of allenes, 132–133
Methylmagnesium bromide, conjugate addition
    of, 144–145
Methyl p-nitrocinnamate, 59
N-Methyl-N-nitroso-p-toluenesulfonamide, 165
  in generation of diazomethane, 144, 155–156
10-Methyl-$\Delta^{1(9)}$-octalone-2, 28, 102–103, 160
Methyl n-octyl ether, 60
Methyl oleate, 163
  deuteration of, 44
4-Methyl-1-pentene, 162
  hydroboration of, 32
Methyl n-pentyl ketone, 95
ω-(Methylsulfinyl)acetophenone, 94
  reduction of, 95
Methylsulfinyl carbanion, 92–95
  in formation of β-hydroxysulfoxides, 93
    of β-ketosulfoxides, 94–95
  in generation of acetylides, 124
  in Wittig reaction, 106–107
Methylsulfinyl cyclohexyl ketone, 94
  reduction of, 15
Methylsulfinyl n-pentyl ketone, 94
  reduction of, 95
Methyl 9,10,12,13-tetradeuterosterate, 44
Methyltriphenylphosphonium bromide, 105, 106
δ-Methyl-δ-valerolactone, 10
Methyl vinyl ketone, 163, 164
  cyclization with 2-methylcyclohexanone, 102
  reaction with enamines, 82–83
    with trialkylboranes, 114
Michael addition, with enamines, 80, 82–85
Mixed hydride reduction, 20–22
Monoperphthalic acid, 154
  in epoxidation of cholesteryl acetate, 89
Morpholine, 163
  in enamine formation, 80–81
1-Morpholino-1-cyclohexene, 80–81, 159, 160,
    163

1-Morpholino-1-cyclohexene—*cont.*
  acylation of, 81–82
  reaction with methyl vinyl ketone, 82–83
    with β-propiolactone, 83–84

N

Naphthalene, reduction of, 25–26
1-Naphthol, 67
2-Naphthol, 67
1,4-Naphthoquinone, 163
  hydrogenation of, 44
NBS, 48–49
Neopentyl alcohol, conversion to chloride, 46
Neopentyl bromide, 47
Neopentyl chloride, 46
Nitriles
  from alkyl chlorides, 140–141
  hydrolysis of to amides, 56–57
Nitroalkanes, carboxylation of, 99
4-Nitrobenzaldehyde, 6
p-Nitrobenzamide, 57
p-Nitrobenzonitrile, hydrolysis, 57
4-Nitrobenzyl alcohol, 6
p-Nitrobromobenzene, by triphenylphosphine
    dibromide, 48
p-Nitrocinnamic acid, 59
  methyl ester, 59
Nitroferrocene, 65–66
p-Nitrophenol, 48
α-Nitropropionic acid, 98
β-Nitrostyrene, 163
  hydrogenation of, 44
Nitrosyl chloride, 11–12, 161
  addition to cyclohexane, 11–12
Nonanamide, 142
1-Nonyne, 124
exo-Norbornane-2-carboxamide, 142
Norbornene, 162, 163, 164
  conversion to dialkylketone, 113
    to trialkylcarbinol, 111–112
  hydroboration, 33
  oxymercuration of, 61, 62
  reaction with formamide, 142
exo-2-Norborneol, 33, 62
exo-2-Norbornyl acetate, 62
Norcarane, 117

O

Octalin, 25–27
  by dehydration of 2-decalol, 56